SCSI

Ulrich Weber

The Ins and Outs

Technology
Devices
Drivers

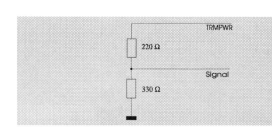

Elektor Electronics (Publishing)

Elektor Electronics (Publishing)
P.O. Box 1414
Dorchester DT2 8YH
England

All rights reserved. No part of this book may be reproduced in any material form, including photocopying or storing in any medium by electronic means and whether or not transiently or incidentally to some other use of this publication, without the written permission of the copyright holder except in accordance with the provisions of the Copyright, Designs and Patents Act 1988 or under the terms of a licence issued by the Copyright Licensing Agency Ltd, 90 Tottenham Court Road, London, England W1P 9HE. Applications for the copyright holder's written permission to reproduce any part of this publication should be addressed to the publishers.

The publishers have used their best efforts in ensuring the correctness of the information contained in this book. They do not assume, and hereby disclaim, any liability to any party for any loss or damage caused by errors or omissions in this book, whether such errors or omissions result from negligence, accident or any other cause.

British Library Cataloguing in Publication Data

A catalogue record for this book is available from the British Library

ISBN 0 905705 44 0

First published in the United Kingdom 1996

© Segment BV

Printed in the Netherlands by Technipress, Culemborg

Contents

Foreword 9

History and development 11
   The Shugart company and the SASI interface 11
   The SCSI concept and its advantages 12
      Updatable 15
   Initiator and Target – an overview 17
      ID numbers 20
      Termination 20
   Practical use – some notes 21
      Common Command Set 22
      SCSI-2 22
      Fast-SCSI and Wide-SCSI 23

SCSI Basics 29
   The control lines 29
      Control lines with Initiator access 29
      Control lines with Target access 31
   Bus phases 32
      Bus-Free Phase 33
      Arbitration Phase 34
      Selection Phase 35
      Reselection Phase 39
      Transfer phases 40
      Message-Out Phase 40
      Message-In Phase 41
      Command Phase 42
      Data-In (-Out) Phase 43
      Synchronous data transmission 44
      Fast-SCSI 45
      Wide-SCSI 47
      Status Phase 49
   The SCSI Protocol 49
      Exception Conditions 50
      Attention Condition 50

    Reset Condition 50
    Nobody to talk to 51
    Message codes 52
    Initiator messages 55
    Abort 55
    AbortTag 55
    Bus Device Request 55
    Clear Queue 56
    Head of Queue Tag 56
    Initiator Detected Error 56
    Message Parity Error 56
    Ordered Queue Tag 56
    Release Recovery 56
    Terminate I/O Process 56
    Target messages 57
    Command Complete 57
    Ignore Wide Residue 57
    Linked Command Complete 57
    Linked Command Complete with Flag 57
    Modify Data Pointer 57
    Restore Pointers 57
    Save Data Pointer 58
    Messages for both directions 58
    Disconnect. 58
    Identify 58
    Initiate Recovery 58
    Message Reject 58
    Simple Queue Tag 59
    Synchronous Data Transfer Request 59
    Wide Data Transfer Request 60
    Structure of the SCSI Target 60
    Tagged queues 61
    Rejections and error reports 62
    I/O operations 63
    Data Pointer 65

The SCSI commands 66
    Command classes 66
    Device classes 67
    Mode parameters 67
    Command building 67

    Command structure 68
    The Status report 72
    Generic commands 73
    Test Unit Ready (opcode 00h) 74
    Inquiry (12h) 75
    Request Sense (03h) 79
    Error feedback 79
    Send diagnostic (1Dh) 87
SCSI device classes 93
    Direct Access Devices 93
    Mapping 93
    Data cache 95
    Removable disks 97
    SCSI commands for Direct Access Devices 98
    Interleave factor 100
    Sequential Access Devices 101
    Physical elements 102
    Partitions 103
    Logic elements 105
    Tapemarks 105
    Blockgaps 106
    Databuffers 106
    Printer devices 108
    Processor devices 109
    CD-ROM devices 111
    Data blocks on the CD-ROM 111
    Optical Memory devices 115
    Write-Once devices 119
    Scanner devices 120
Media-Changer devices 123
    Communication devices 124

SCSI in practice 129

Configuration of the SCSI bus 129
    PCI or what? 130
    Termination 132
    Active terminators 136
    ID numbers 138
    Priority assignment 139

Parity checking	139
SCSI adapter installation	142
SCSI cables	143
Internal SCSI cables	144
External SCSI cables	146
Cable quality under Fast-SCSI	151
Cable standards for SCSI-3	152
Plug connections	152
Software interfacing	154
CAM (Common Access Method)	155
Device drivers	156
XPT functions	157
SIM functions	157
ASPI (Advanced SCSI Programming Interface)	158
SCSI BIOS	160
SCSI BIOS boot helper	163
Int 13h	163
SCSI adapters and their installation	165
General installation notes	165
PCI requirements	167
Symbios 8250S	168
Drivers for DOS and Windows 3.x	170
BIOS version 3.x	170
CAM drivers	171
MINICAM.SYS	172
DOSCAM.SYS	172
Device driver	173
SCSIDISK.SYS	174
CDROM.SYS	175
Drive letter allocation	176
Multiple CD-ROM drives	177
ASPICAM.SYS	177
BIOS version 4.0x	178
Error causes	181
Drivers for Windows 95	182
Drivers for OS/2	185
Drivers for Windows NT (3.51)	186
Drivers for Netware 3.1x and 4.x	187
Drivers for UNIX	189

SCO UNIX	189
UnixWare 1.x/1.2	189
Adaptec AHA-2920	190
EZ-SCSI 4.0	191
The installation	192
Wide-SCSI adapters	195
Comparing figures	198
32-bit Wide and Ultra-SCSI	200
ID numbers	201
Termination	201
Meaningful distribution of data	203
RAID systems	204
Disconnect/Reselect with hard disks	205
AHA-2940 UW	207
Plug & Play	209
SCAM	210
SCSI devices	214
Internal or external devices?	214
Magnetic disk drives	216
Large hard disks	221
Removable disk drives	222
CD-ROM drives	223
CD changers	227
CD writers	228
PD drives	229
MO drives	231
Tape drives	233
Scanners	236
Twain	237
Digital cameras	238
External SCSI adapters	239
SCSI and PCMCIA	239
SCSI via the printer port	242
SCSI-3 — an outlook	**245**
Modular structure	245
Fibre Channel	249
FC-AL	252

SSA 253
    Fire Wire 253
    Parallel or serial? 254

Troubleshooting guide **255**

Appendix **267**
Differential SCSI 267
    Signal levels 268
SCSI cable connector pinouts 271
The Companion CD-ROM 277
Glossary 279
Index 287

Foreword

SCSI, the *Small Computer System Interface*, has been with us for nearly fifteen years now, and was constantly enhanced over this period. Today, it is considered to be the most powerful and up-to-date bus system for the connection of peripheral devices, running on widely different computer platforms.

A well-established standard on Apple computers and other systems, SCSI was long branded a technically advanced but expensive system when it came to application within the world of Intel/Windows PCs. But there, too, the arrival of new operating systems (OS/2, Windows 95, Windows NT) has changed opinions as a result of DOS and Windows 3.1 being slowly pushed in the background, facing a growing market share for operating systems capable of exploiting the functionality of SCSI to the full.

The theory section in this book provides readers with background knowledge which is necessary to be able to understand the operation of the SCSI system and, if necessary, program their own drivers using the deepening information found on the companion CD-ROM.

The chapter devoted to practical aspects of the system provides an extensive description of the configuration of the SCSI bus, and supplies tips, suggestions and assistance with problematic cases. The information quickly provides you with an overview regarding the choice of SCSI components which best suit your needs, and getting the best possible performance out of your system.

I hope that readers contemplating the use of a SCSI system for the first time, as well as practically-minded user wishing to extend their knowledge find the book a comprehensive source of information concerning the SCSI system.

Aachen, October 1996

Ulrich Weber

1. History and development

Today, the SCSI interface is available on almost any computer system of the professional class, for the connection of equipment which may be classified as memory storage devices in the widest sense of the word. Practically all devices, from exotic one-offs to hard disks with sales volumes running into hundreds of thousands, may be operated on one and the same bus system. Driver problems are few and far between, and incompatibility between different devices unheard of.

That was not always the case, however. Until the universal and unproblematic use of SCSI devices was reached, the going was rough initially, as with most new technology.

When the acronym SCSI (Small Computer System Interface) was elected as the name for a new I/O standard, that event did not necessarily mark the birth of a new peripheral bus system – the case had a history.

1.1 The Shugart company and the SASI interface

The history goes back to the year 1979 when American hard disk manufacturer Shugart was pondering over a new interface for his hard disk drives, and summarized the results of the considerations under the name SASI (Shugart Associates Systems Interface).
The reason for a manufacturer of magnetic disk drive units to start thinking about a new interface concept was, basically, the following:

Up to then, only direct addressing of drive blocks was known. The SASI interface however would allow, for the first time, so-called logic blocks to be addressed. The disadvantage of the previous addressing method was that the computer had to be taught, over and over again, the proper way to communicate with a new peripheral using different block sizes, any time one was added to a system. This instructing of the computer was achieved by means of a new hardware component (insertion card) and/or a new software module (driver). Because the development of the PC was in its infancy at that time, technical novelties popped up frequently in the area of peripheral equipment. Soon, the adaptation of drivers and insertion cards lost track of this much faster development. No sooner had communication been achieved between a computer and a peripheral, or a further development of the peripheral appeared, which was as yet unusable to the computer. The actually usable peripheral devices were always one step behind the latest developments.

SASI, it was hoped, would eliminate the discrepancy between the technically possible and the actually usable. But it was not just the novel initiative that would cause SASI to become the origin of the SCSI standard. Shugart, by publicizing the specifications of his interface, created the basic requirement that would allow the further development of his idea, under the name SCSI, to evolve into the ANSI (American National Standards Institute) standard.

The ANSI working group met for the first time in 1982. Its aim was develop a new, powerful bus system for peripheral data carriers, based on the SASI interface. The project received the name SCSI for *Small Computer System Interface*. The leading manufacturer at the time was NCR.

Two years later, the ANSI committee was presented a document under the name SCSI Revision 14. After two more years, in June 1986, the goal was reached – SCSI Rev. 17b was declared an ANSI standard.

1. 2 The SCSI concept and its advantages

In the years 'Before SCSI', the computer had to be taught how to communicate any time a newly developed peripheral was connected up. In practice, that meant at least one adaptation at the software level for each new device. In many cases, a hardware solution had to be found, too, so as to enable the computer and the peripheral to talk to each other. Such equipment adaptations took a lot of time; time in which peripheral devices were developed out. No matter how much effort was put into develop-

1.2 The SCSI concept and its advantages

ment, equipment adapting was always behind the development of new peripheral equipment. The instant the computer had learned to communicate with a certain peripheral, the development of the same peripheral was a step ahead. Peripherals that were actually usable within a certain system were outdated by definition.

SCSI opened new possibilities in this respect. As a device independent I/O bus, the SCSI system is built such that the processor need not bother to know the exact type of the equipment which is hooked up to the bus. That is because SCSI hides the structure of the connected peripheral for the computer. Each apparatus appears on the bus as a device with a certain number of logic blocks. Because a hardware-based (physical) examination of the peripheral is not carried out, the drive system is considerably simplified. In addition, a connected peripheral is allowed to provide indications on its (logic) size and device class. Arguably, that also lightens the use of drivers which are size and manufacturer independent.

A SCSI *interface* is used to link a peripheral device to a SCSI bus. This interface communicates with the SCSI bus on the one side, and on the other side, with the peripheral. It is normally fitted in the peripheral device, which has helped to frame the very denomination *SCSI devices*. The task of a SCSI interface is to convert the device-specific signals in such a way that they comply with the SCSI standard, and may be injected into the SCSI bus system.

All equipment having a SCSI interface follows the same standard to put signals on to the bus. Consequently, such equipment may also interchange data, simply because they 'speak the same language'.

The SCSI bus is coupled to a processor by means of a so-called SCSI adapter. The relations are illustrated in Figure 1.1.

The denotations used in professional literature can easily cause confusion in this context. Originally, the hardware link between computer and SCSI bus was called *SCSI adapter*, while the link between the SCSI bus and the peripheral was referred to as a *SCSI controller*. With other interfaces like IDE and others, the word 'controller' is, however, fairly common for the insertion card which forms the connection with the processor. Because of this, the hard-

1. History and development

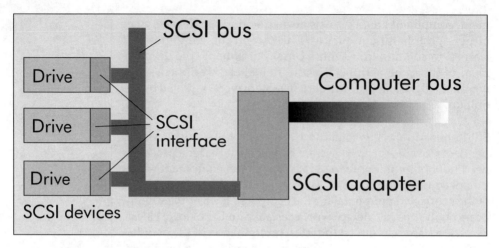

Figure 1.1.
Basic structure of a SCSI system.

ware bridge between the SCSI bus and the peripheral is currently called *SCSI interface* to prevent confusion with the SCSI adapter.

To recapitulate, the above mentioned terms are explained once more:

- **SCSI adapter:** hardware bridge between the SCSI bus and the computer bus (ISA, EISA or PCI with IBM compatible PCs, NuBus or PCI with Apple computers). The SCSI adapter is either available as a plug-in card, or as an integral circuit on the motherboard of the computer.

- **SCSI interface** (or SCSI controller): a manufacturer-specific unit, which forms the link between the peripheral device and the SCSI bus. It is usually housed in the peripheral equipment. The type and size of the SCSI interface are governed by the manufacturer, and determine to a large extent the abilities of the relevant device.

Having explained the terms SCSI adapter and SCSI interface, the question remains how many devices may be operated on a SCSI bus.

1. 2 The SCSI concept and its advantages

Updatable

Before answering this question, a short excursion. As already mentioned, the workgroup report designated SCSI Rev. 17b was raised to a standard by the ANSI committee. That did not, however, mean that the work of the SCSI workgroup was finished. On the contrary, SCSI had to prove itself in practice. A rather large margin was allowed for interpretation of the standard by equipment manufacturers. Too large, it appeared later. None the less, the workgroup was in a position to learn from errors and obstacles. SCSI-2 was pulled from the hat, its main point being the integration of a *Common Command Set* (CCS) for uniform driving of hard disks, while Fast-SCSI and Wide-SCSI were laid down as extensions of SCSI-2. All completing functions and extensions have one thing in common: they are downward-compatible. Errors were eliminated and necessary additions incorporated. The advancing technology was taken into account, but the extension of the system took place without irregularities. Although all extensions will be discussed in detail in this book, I would like to describe them as they are: extensions, which are result of continuous growth.

The question of the maximum number of SCSI devices is answered differently in relation to Wide-SCSI than in relation to the 'normal' SCSI bus, because the maximum number depends on the number of available data lines. All values mentioned in this introductory chapter therefore refer to 'normal' SCSI, unless otherwise noted.

The ANSI recommendation allows a maximum of 8 devices to be connected to a SCSI bus. This number should include at least one SCSI adapter, because the system can not operate without a link to a processor. If required, however, more devices are allowed. There are two permissible extremes: seven SCSI adapters and one SCSI device (a rare configuration by almost any standard), and one SCSI adapter and up to seven SCSI devices. Each solution in between these bounds is also allowed. The minimum configuration consists of one SCSI adapter and one SCSI device. The relations are clarified by Figure 1.2.

The standard configuration will look like this in most PCs: one SCSI adapter and up to seven SCSI devices which transmit data to

1. History and development

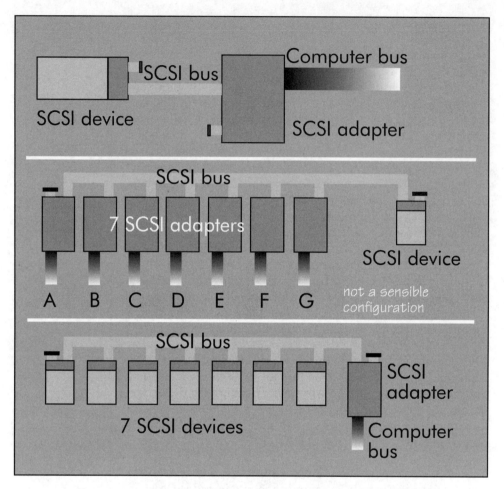

Figure 1.2.
Minimum and maximum configuration on an 8-bit wide bus.

the adapter, or receive data from it. The Host adapter is, therefore, a common junction to the computer bus for all SCSI devices. Consequently, SCSI is not only *the* communication link between the various peripheral devices and the computer; a single host adapter is also capable of serving up to seven devices, where other interfacing methods, in the worst case, require one insertion card per device.

Apart from the usual lack of space (who has seven free slots in his computer?), resource problems also occur. At least in IBM compatible PCs, each insertion card preferably has its own interrupt, while cards with just about any function related to data transfer require a free DMA channel, too.

Whether or not the right device interface is available for the desired computer bus also remains to be seen – PCI or perhaps only ISA?

These problems do not exist under the SCSI standard. Host adapters are available for practically any computer-internal bus type. In case the SCSI adapter is not already integrated on the motherboard, then all you have to do is install a plug-in card (the SCSI adapter). The connection of the SCSI devices is then carried out in *SCSI-internal* fashion.

1.3 Initiator and Target – an overview

On a SCSI bus, no distinction is made primarily between a SCSI device and a SCSI adapter. Instead, a separation is made between driving and driven element, or *Initiator* and *Target* respectively. The Initiator gains access to the bus and selects the Target it wants to communicate with. Once this selection procedure is finished, the Target arranges the driving of the data exchange to and from the Initiator. Because it is possible for more than one SCSI adapter to operate on the bus, a SCSI adapter may be an Initiator as well as a Target.

To obtain an initial impression of the basic function sequences in a SCSI system, it is necessary to first examine the structure of the SCSI bus in some detail. The SCSI bus consists of 50 lines (wires) of which all odd-numbered lines (1, 3, ... 49) as well as lines 20, 22, 24, 28, 30 and 34 are at ground potential, and pin 25 is not used. Table 1.1 shows that nine of the remaining lines are reserved for data transport (eight datalines and one parity check line, D0 - DBP), and another nine lines (ATN through I/O) are used for control functions. Pin 26 (TERMPWR) carries a 5-V direct voltage which may be used to power Terminator devices.

1. History and development

The eight-bit wide databus allows a maximum data rate of 5 MByte/s to be reached, depending, however, on different operating modes on the bus. Although only one signal mode is customary on the SCSI bus, and this mode is internally driven, all available options are mentioned below for the sake of completeness.

With SCSI a distinction is made between

- single-ended signals
- differential signals

and

- asynchronous data transfer
- synchronous data transfer

SCSI supports both, mutually incompatible, signal forms: single-ended and differential. In the single-ended version, TTL levels are used (the signal level swings between a defined positive extreme and ground), while in differential mode positive and negative levels are defined (EIA RS-485 signal). Although the cables used are identical, the pin assignment differs between these versions. Table 1.1 indicates the contact assignment of the single-ended version. To prevent confusion, the same table for the exotic differential mode is not shown here but in an Appendix.

Both signal forms are mutually incompatible, and devices of both versions may not be operated on the same bus. The advantage of differential SCSI is a larger allowable cable length (25 m as opposed to 6 m for the single-ended version). This may be useful when SCSI apparatus in different rooms are integrated into a system.

Consequently, the differential SCSI version is only found on large server systems, and not even frequently in these surroundings. Differential SCSI devices are few and far between on the market, expensive, and really without significance for the average PC user. The main part of this book therefore concentrates on single-ended SCSI. Differential SCSI is discussed separately in an Appendix.

1.3 Initiator and Target – an overview

With the transfer modes, things look rather different; although both are available on the bus, that circumstance need not worry the user. The terms synchronous and asynchronous transfer mode refer to the way in which handshaking, that is, the acknowledging of received signals, is dealt with. Asynchronous mode requires handshaking after each received signal. In synchronous mode, a series of signals may be conveyed before a handshake should occur. Consequently, synchronous data transfer involves a smaller handshaking overhead, which boosts the data transfer speed: up to 5 MByte/s may be achieved in synchronous mode, while asynchronous data transfer is limited to a maximum speed of about 1.5 MByte/s.

Pin	Signal	Description
2	-DB0	Data Bus Line 0
4	-DB1	Data Bus Line 1
6	-DB2	Data Bus Line 2
8	-DB3	Data Bus Line 3
10	-DB4	Data Bus Line 4
12	-DB5	Data Bus Line 5
14	-DB6	Data Bus Line 6
16	-DB7	Data Bus Line 7
18	-DBP	Data Bus Parity
26	TERMPWR	Terminator Power
32	-ATN	Attention
36	-BSY	Busy
38	-ACK	Acknowledge
40	-RST	Reset
42	-MSG	Message
44	-SEL	Select
46	-C/D	Control/Data
48	-REQ	Request
50	-I/O	Input/Output

Table 1.1.
Pinout of the 50-way SCSI cable (single-ended, 8-bit). All odd-numbered connections, as well as pins 20, 22, 24, 28, 30 and 34 are at ground potential.

The asynchronous transfer mode is the 'default' mode on the SCSI bus, and the bus is automatically set to function in this mode after a computer reset or a reset of the SCSI system. All commands, status reports, etc., are always transmitted in asynchronous mode. If a SCSI device is connected which is able to communicate in synchronous mode, the Initiator and the Target may agree on using this mode for their **data** exchange operations (provided, of course, both are capable of using synchronous transfer). Details on the exact operation of this option are given further on. Here, however, I should add that these 'agreements' on the bus are so far-reaching that neither the SCSI adapter nor the CPU may be involved in a data exchange between two SCSI devices. Consequently, the CPU may turn to other tasks during a copy process, provided, of course, that is made possible the operating system.

The individual phases which are significant on the SCSI bus during a data transfer operation are discussed in detail in Chapter 2 of this book.

ID numbers

Interested readers will no doubt have wondered by now how several SCSI devices on the bus may be kept apart and addressed separately. After all, it was already mentioned that the SCSI system does not really bother to be aware of the exact type of SCSI device it is communicating with.

On an eight-bit wide SCSI bus, only eight SCSI devices may be used, including the adapter. So-called ID numbers are used for addressing. These numbers range from 0 to 7, and their allocation determines the priority sequence on the bus. Number 7 has the highest priority, and is usually reserved for the SCSI adapter of the host computer. So, the up to seven SCSI devices have numbers 0 through 6 to choose from. These numbers are usually allocated by the user as he or she connects the devices. Care should be taken, of course, to prevent a certain number being allocated twice. Within the framework of Plug And Play, there are initiatives which indicate that newly arriving devices are increasingly capable of setting their own address to a free ID number. Although this helps to avoid duplicate ID numbers, it solves the matter of device priority in a rather random manner.

Termination

The second adjustment the SCSI user has to make, in as far as he or she is not helped by automatic Plug And Play setup, is the bus termination. Without preempting a detailed discussion in the practical part of this book, the following is noted for a general understanding:

The SCSI bus must **always** be terminated with a terminator at both ends. In the simplest case, the terminator is a resistor array which guarantees defined signal levels on the SCSI bus lines.

1. 4 Practical use – some notes

If a SCSI adapter of the insertion card type is applied, as illustrated in Fig. 1.3, there is usually an external and an internal SCSI connection. The external connection is used to connect SCSI devices outside the computer, while the internal connection is provided for devices within the computer enclosure. If both internal and external devices are connected, terminators are fitted at each end, on the last device. If one side is not used, the non-used cable is still terminated at the SCSI adapter.

1. 4 Practical use – some notes

The foregoing should provide you with a short and superficial glance at the operation of the SCSI system. The introduction is adequate, however, to be able to confirm the strong and weak sides of SCSI (-1), as well as understand the improvements to the system which arrived by and by during its development.

SCSI-1 was originally announced as a new, extremely powerful, interface for peripheral data carriers ('media'), aimed at eliminating the driver problems with newly developed peripherals, as ex-

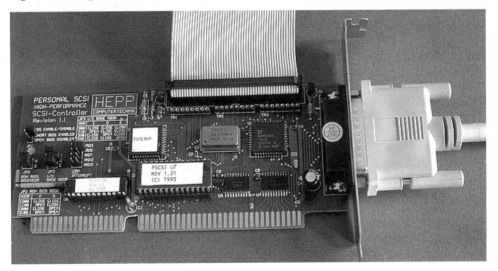

Figure 1.3.
External (right) and internal SCSI cable on a SCSI adapter.

plained at the start of this chapter. The connection of a SCSI device would be reduced to connecting a cable and setting the right ID number. The task of all equipment manufacturers, then, was to adapt their different products via a device-specific interface, in such a way that the command set made available by SCSI would enable all existing and future peripherals to be used without problems on the SCSI bus.

As regards its performance (data transfer rate, CPU load, etc.), SCSI-1 obviously satisfied the expectations. By contrast, the notorious driver problems did not cease to occur, and SCSI was unsatisfactory in this respect.

Many functions and commands of the SCSI-1 standard were either only optional, or not defined at all. The amount of freedom given to device manufacturers with the aim of being able to respond quickly to technical novelties, turned out to be a serious disadvantage.

Because SCSI addresses logic blocks, it is necessary to provide, for example, commands that allow hard disks to be formatted, or data on defective tracks to be restored. This area was, however, entirely left to the gusto of the hard disk manufacturers, so that each of these came up with its own solution. As a result, each of these solutions required a specific driver, and as regards compatibility users were back to square one.

Common Command Set

The problems caused by SCSI were not only recognised by the users (*'Apparently SCSI and SCSI are not always the same'*), but also by the manufacturers, who found themselves more or less forced to combat incompatibility problems. The leading hard disk manufacturers agreed upon a **Common Command Set** (CCS), a collection of 18 sub-commands which were to be integrated into the SCSI standard, and were aimed at providing a guide to hard disk manufacturers. CCS was offered to the ANSI committee as a proposition.

SCSI-2

If CCS was a first step in the desired direction, then SCSI-2 completed all further steps which were required. All options that could cause incompatibility were removed. Eleven device classes were introduced, and CCS defined for these, too. Within a device class, each device is treated equally. In combination with the addressing of logic blocks, that approach helps to simplify the work of the software developer. The problem of the best possible matching of a specific device is now moved to the device side, and solved by the SCSI interface. Thus, the device manufacturer can no longer shift device adaptation to an additional piece of driver software.

SCSI-2 actually managed to reach the goal of being an equipment-independent software interface.

Fast-SCSI and Wide-SCSI

Fast-SCSI and Wide-SCSI are basically extensions of SCSI-2 that allow a much higher data exchange rate to be achieved.

Fast-SCSI is an optional version of the regular 8-bit SCSI. The synchronous data transfer rate is raised from 5 to a maximum of 10 MBit/s by shortening the signal times on the REQ and ACK control lines (see Chapter 2). To be able to actually utilize the higher data transfer rate, the SCSI adapter and at least one connected device should be Fast-SCSI compatible. Normal SCSI-2 devices may be used on the same bus, using their normal, lower, data transfer speed.

With Fast-SCSI, the maximum cable length is reduced to 3 m.

Wide-SCSI boosts the data throughput by doubling (or quadrupling) the number of datalines. Assuming that the Wide-SCSI adapter actually employs the improved signal protocol of Fast-SCSI, data transfer rates of up to 40 Mbyte/s may be achieved.

As a matter of course, different cable and connector types are used with 16- and 32-bit wide data buses. The currently established 16-bit Wide-SCSI adapters do, however, allow 8-bit as well as 16-bit devices to be employed simultaneously (A- and P-Cable) at the internal side.

1. History and development

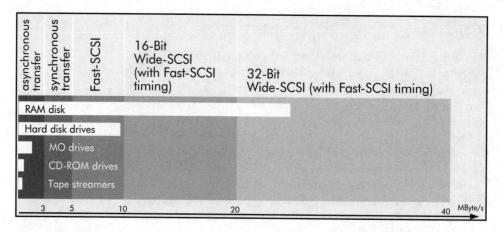

Figure 1.4.
The data exchange rates that may be achieved with the different SCSI versions as compared with the transfer rates managed by commonly used SCSI devices.

Wide-SCSI allows up to 32 SCSI devices to be connected (including the Host adapter).

Meanwhile, its excellent compatibility and performance have made SCSI an industry standard for the connection of peripheral data carriers on nearly all of today's computing platforms. The possibility of being able to use SCSI equipment with widely differing operating systems and hardware configurations creates a high degree of investment security, while prices of SCSI devices are coming down rapidly thanks as a result of increased use by private computer owners.

The fact that the SCSI standard is open ensures a continuous adaptation to new developments (SCSI-3, Ultra-SCSI, etc.), while SCSI is also certain to remain what it is today:

The standard for connecting peripheral devices right across all computing platforms.

1.4 Practical use – some notes

- Restricted by the addressing of logic blocks, widely differing peripheral devices may be connected to the SCSI bus.
- SCSI-2 guarantees that these devices, different as they may be, may be utilized without driver problems.
- The SCSI system offers very high data transfer rates, as well as sufficient spare power for future technology.
- Nearly all control and checking functions are carried out by the SCSI units, lightening the CPU load.
- A single host adapter enables up to seven SCSI devices (or 15 devices with Wide-SCSI) to be connected, that is, one insertion card for 7 (15) devices.
- SCSI may be employed with all major computer bus systems. Consequently, peripheral devices may continue to be used after a computer upgrade.
- It is also possible to fit an 'internal' device in its own enclosure, allowing it to be used with different computers without hardware conversion work.
- SCSI is constantly developed further, ensuring new technology is duly taken into account.

Table 1.2.
Main advantages of the SCSI system.

Small Computer System Interface

Theory

- Control Lines
- Bus Phases
- Transmission Modes
- SCSI Protocol
- SCSI Commands
- Devices Classes

2. SCSI Basics

As opposed to the introductory overview the first chapter is intended to provide, the following descriptions and illustrations form a step-by-step introduction into the functional operation and control of the SCSI bus. As already mentioned, all descriptions will relate to *single-ended SCSI*. Details on differential SCSI may be found in an Appendix.

2.1 The control lines

The standard SCSI bus (called 8-bit SCSI bus in the following discussion) provides nine datalines (including parity checking), nine control lines, and one line marked TERMPWR (*Termination Power*) which supplies 5 V to the terminators. The pinout of the bus connector was already shown in Table 1.1. For a better overview, the signal/cable allocation is repeated here, and completed with the column *access control* indicating which type of device is allowed to use a particular line (Table 2.1). The remaining 31 wires of the 50-way cable are connected to ground.

Control lines with Initiator access

As already mentioned, the primary distinction on a SCSI bus is not between SCSI adapter and SCSI device, but between Initiator and Target. As opposed to other I/O systems using a fixed Master/Slave assignment, the allocation between Initiator (Master) and Target (slave) may be swapped. So, a SCSI adapter need not always be an Initiator, while a Target may also be called upon to perform control functions (Master functions) in many sequences. In this context, the SCSI bus is said to have Multi-Master Ability.

Consequently, the control lines on the bus are used by Initiators as well as Targets to control and monitor processes on the bus.

2. SCSI Basics

signal name	ext. plug	wiring and int. plug		ext. plug	signal name	access control
GROUND	1	1	2	26	-DB(0)	
GROUND	2	3	4	27	-DB(1)	
GROUND	3	5	6	28	-DB(2)	
GROUND	4	7	8	29	-DB(3)	
GROUND	5	9	10	30	-DB(4)	
GROUND	6	11	12	31	-DB(5)	
GROUND	7	13	14	32	-DB(6)	
GROUND	8	15	16	33	-DB(7)	
GROUND	9	17	18	34	-DB(P)	
GROUND	10	19	20	35	GROUND	
GROUND	11	21	22	36	GROUND	
RESERVED	12	23	24	37	RESERVED	
FREE	13	25	26	38	TERMPWR	
RESERVED	14	27	28	39	RESERVED	
GROUND	15	29	30	40	GROUND	
GROUND	16	31	32	41	-ATN	Initiator
GROUND	17	33	34	42	GROUND	
GROUND	18	35	36	43	-BSY	Initiator / Target
GROUND	19	37	38	44	-ACK	Initiator
GROUND	20	39	40	45	-RST	Initiator / Target
GROUND	21	41	42	46	-MSG	Target
GROUND	22	43	44	47	-SEL	Initiator / Target
GROUND	23	45	46	48	-C/D	Target
GROUND	24	47	48	49	-REQ	Target
GROUND	25	49	50	50	-I/O	Target

The (-) character in front of signal and data lines indicates that the LOW level is actively controlled with all SCSI devices while a HIGH level marks a passive state.

Table 2.1.

Signal/cable assignment on the 50-way SCSI cable (single ended) with indications regarding access permission for Initiator and Target.

However, to do so neither Initiator nor Target have available all nine control lines.

An Initiator can only make use of the following five control lines:

- RST – The RST signal (*Reset*) enables the Initiator to reset the bus.
- BSY – The BSY (*Busy*) is used by the Initiator to state that the bus is occupied.
- SEL – The SEL (*Select*) signal allows the Initiator to select a Target.
- ACK – Using the ACK (*Acknowledge*) signal, the Initiator confirms a request received from the Target.
- ATN – The Initiator uses the ATN (*Attention*) signal to tell the Target that it wants to send a message.

Control lines with Target access

A Target is able to use seven of the nine control lines. Three of these are shared with the Initiator, or any other device; four control lines are exclusively reserved for Targets (see Table 2.1).

- RST – Using the RST signal (*Reset*) a Target is able to reset the bus.
- BSY – A Target may use the BSY signal (*Busy*) to declare the bus occupied.
- C/D – The C/D (*Control/Data*) signal is used by the Target to indicate the difference between control signals and data. An active signal level indicates that a control signal is involved.
- I/O – The I/O signal (*Input/Output*) is employed by the Target to define the data transfer direction with respect to the Initiator. An active signal level indicates that the Initiator reads data.
- MSG – The Target may use the MSG (*Message*) signal to inform the Initiator that a message will be conveyed.
- REQ – During the handshaking process, a REQ (*Request*) signal issued by a Target prompts the Initiator to start the data exchange.
- SEL – The SEL (*Select*) enables a Target to return the control of the bus to an Initiator.

This set of control signals characterizes as well as controls the different states on the SCSI bus, which are known as *Phases*.

The datalines DB0 through DB7 may be set by Initiator as well as by Target devices. This feature is significant with the selection of the desired SCSI devices (ID number).

2. 2 Bus phases

The SCSI bus has eight different phases, which are determined by control signals. The phases determine in which way data is to be conveyed over the datalines, and in which direction.

Each operation on the SCSI bus starts with the *Bus-Free Phase*, and is also ended with it. The SCSI bus is also in this state after a processor restart, or a SCSI reset.

An overview of the Bus Phases:

- Bus-Free Phase   Quiescent state of the bus, no device is active on the bus, all signals are at logic 0 (inactive).
- Arbitration Phase   The allocation phase; in this phase, the Initiator with the highest priority is awarded control of the bus.
- Selection Phase   During this phase an Initiator selects a Target.
- Reselection Phase   During this phase, a Target which was temporarily disconnected from the Initiator, re-establishes the original connection.
- Command Phase   During this phase, commands are conveyed.
- Data Phase   Depending on the direction of the data transfer, the term *Data-In* or *Data-Out* is used. This is the only phase during which synchronous data transfer may be set up instead of asynchronous.
- Status Phase   During this phase, the status byte is normally copied to the Initiator.

- Message Phase   This phase is used for the connection and phase control. Depending on the direction, a distinction is made between *Message-In* and *Message-Out*.

The sequence of the individual bus phases is shown is a simplified way in Fig. 2.1.

Each phase may be ended by a *Reset* command, after which the bus always goes into the *Bus-Free Phase*. This is also clarified by Fig. 2.1; a *Bus-Free Phase* may follow on any SCSI phase.

For the sake of completeness, it should be mentioned that there were, and still exist, systems without *Arbitration Phase* under SCSI-1. The structure of the relevant SCSI adapters and associated drivers is such that communication is possible with one SCSI device only. In this context, I like to speak of 'narrow-track adapters'. These adapters are often supplied with scanners and CD-ROM players. They are inexpensive (not cost-effective), and claimed to ensure the operation of the associated hardware when a full-blown SCSI adapter is not available. Because such an adapter allows only one SCSI device to be connected (normally, the equipment the adapter comes with), all advantages of the SCSI system are lost. In such a system, the sequence of the individual SCSI phases is identical with that shown in Fig. 2.1, only the *Arbitration Phase* is missing because the adapter is unable to select and address different SCSI devices. Further on in this book I will revert to these 'narrow-track adapters'.

Before doing so, however, a closer examination is presented of the individual bus phases. Also described are the signal sequences and the signals on the various control lines.

Bus-Free Phase

The Bus-Free Phase indicates that no SCSI device is active on the bus. This state is reached provided the control signals BSY, SEL and RST are inactive, i.e., at logic 0. All other signals are also at 0 after a delay.

2. SCSI Basics

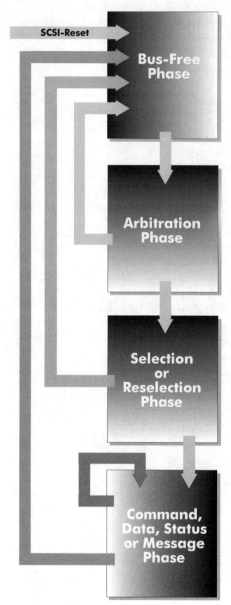

Figure 2.1.
General overview of the sequence and possible transitions between the individual SCSI phases.

In normal, non-interrupted mode, two situations may occur that cause the SCSI bus to go into the Bus-Free Phase:

a. On finishing the Command Phase, after the message *Command Complete* has been conveyed (see Fig. 2.1);

b. After a Target has freed the bus again with the aid of the *Disconnect* message.

In a number of exceptional cases, the Initiator may use certain messages to force the bus to be freed. The meaning of these messages will be reverted to in greater detail, further on.

If an Initiator recognizes a Bus-Free state which was not forced by itself, or brought about by either of the two above mentioned Target messages, it treats the state as an error.

Arbitration Phase

In the Arbitration Phase which follows the *Bus-Free Phase*, each of the SCSI devices may request access to the bus. If more than one device requests the bus, this is allocated on the basis of a priority order.

The following pattern applies if a SCSI device wants to select a Target, or re-establish the connection with an Initiator.

- The device must first obtain permission to access the bus. For this, it awaits the Bus-Free Phase.

- As soon as this phase is recognized, the device, after a short delay, sets the *BSY signal* (*Busy*), and pulls the line corresponding with its own *ID number* to '1'.

- After a short pause (*arbitration pause*), the device checks whether it has the highest priority (*ID number*) among the other devices which are also connected to the bus. If that is the case, the sequence continues with the *Selection* or *Reselection* Phase.

- The SCSI device which has gained access to the bus on the basis of the highest priority sets the SEL signal and waits two time slots (*bus clear delay + bus settle delay*) before it changes one or more signals.

The SEL signal from the device with the highest priority is an indicator to all devices involved in the Arbitration Phase that their own BSY signal should be taken from the bus within a period called *bus clear delay* (approx. 800 ns).

The process of bus allocation as described here is also carried out under SCSI-2 when there is only one Initiator and one Target. Under SCSI-1, this need not be so (see 'narrow-track adapters').

Selection Phase

The Selection Phase is used by the Initiator that has obtained access to the bus to establish the connection with the Target to be addressed. The term *Reselection Phase* is used when a Target wants to re-establish the connection with an Initiator from which it cut off itself. The difference between Selection and Reselection phase is mainly that the I/O line is also set to '1' when the connection is re-established by the Target.

Whether a SCSI device is an Initiator or a Target may be detected by checking for a '0' or a '1' during the Selection Phase.

At the start of the Selection Phase, the BSY and SEL lines used during the Arbitration Phase are at '1', as well as the dataline which corresponds to the Initiator ID (DBn). The Initiator then also sets

2. SCSI Basics

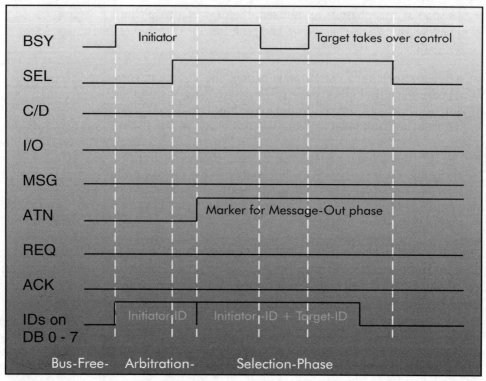

Figure 2.2.
Timing diagram of the Arbitration and Selection phase.

a '1' on the line which identifies the Target ID, and so selects the desired Target.

In addition, the ATN signal is set which marks that the Message-Out Phase follows the Selection Phase. This procedure is prescribed by the SCSI protocol (see Fig. 2.3). The setting of the ATN signal is followed by a defined delay (min. 2 *de-skew delays*), after which the Initiator resets the BSY signal ('0').

In this way, it supplies the start signal for each device on the bus to check the datalines for its own ID number. After a maximum response time of 200 ms (*Selection Abort Time*), the addressed Target is required to respond and take over the control of the bus. This is achieved by the Target setting the BSY signal. As soon as the

2. 2 Bus phases

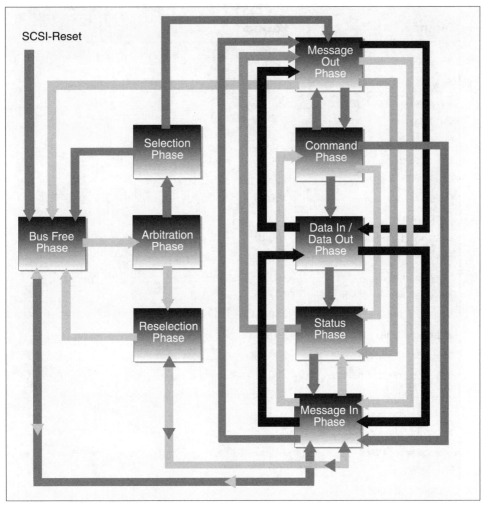

Bild 2.3.
Complete phase diagram of the SCSI bus. All phase sequences allowed by the SCSI protocol are shown.

Initiator detects the BSY signal from the Target, it must reset the SEL signal and the datalines (allowable delay: 2 de-skew delays). From this moment on, the Target is in control of the SCSI bus, until the end of the complete sequence.

2. SCSI Basics

Name	Period
Arbitration delay	2.4 µs
Assertion period	90 ns
Bus clear delay	800 ns
Bus free delay	800 ns
Bus set delay	1.8 µs
Bus settle delay	400 ns
Cable skew delay	10 ns
Data release delay	400 ns
Deskew delay	45 ns
Disconnection delay	200 µs
Hold time	45 ns
Negation period	90 ns
Power-on to selection time	10 s recommended
Reset to selection time	250 ms recommended
Reset hold time	25 µs
Selection abort time	200 µs
Selection time-out delay	250 ms recommended
Transfer period	negotiated during Message-Phase
Fast assertion period	30 ns
Fast cable skew delay	5 ns
Fast deskew delay	20 ns
Fast hold time	10 ns
Fast negation period	30 ns

Table 2.2.
SCSI timing — an overview of the defined delay, pulse-stable and reaction times.

Figure 2.2 shows a signal level diagram of all control lines and data lines during the Arbitration and Selection Phase. At the start of the Arbitration Phase, the control lines *BSY* and *SEL* are set to '1', along with (in this example) dataline *DB7*. In other words, the SCSI adapter (ID number 7) takes over the control of the bus. The start of the Selection Phase coincides with the instant the ATN level goes to '1' while, in addition, a dataline is set to '1' (in this example, *DB2*). Consequently, the Initiator addresses the Target with ID number 2.

At the start and the end of each selection procedure, a short voltage pulse may be noticed on non-used datalines. These so-called *glitches* are actually the reason for the delay periods specified in the SCSI protocol. In general, signal level transitions are hardly ever as clean and rectangular-shaped as suggested by the usual timing diagrams. In practice, signal edges are often much slower or even rounded, while light oscillations (ripple) may even be observed in very bad situations. The quality of the signal edges is affected by the cable in particular. The effect of cable impedance, cross-talk suppression and, in particular, cable length on the edge steepness should not be underestimated.

Because there is no fixed bus clock in the SCSI system, signals are not read at predefined instants. Consequently, *delays* must be defined and maintained to make sure that the signal levels reach their final values before they are processed.

Because a bus clock is not available, another condition must be met to guarantee error-free reading of signal levels:

Each signal must be put on the relevant line with a defined minimum idle time, else, errors can not be avoided. As a result, an accurate timing has to be defined for the various signals and the delays to be maintained (see Table 2.2).

Reselection Phase

During the Reselection Phase, a Target can re-establish the connection with an Initiator, after having freed the bus in the mean time. Beforehand, the Target must have obtained access to the bus during the Arbitration Phase. Once that has been accomplished, it sets the I/O signal, and so identifies itself as a Target. All other processes are identical with the Selection Phase. After the Reselection Phase, an already existing connection is re-established which was broken in the mean time.

2. SCSI Basics

Phase	MSG	C/D	I/O	Direction
Data-Out	—	—	—	Initiator → Target
Data-In	—	—	1	Target → Initiator
Command	—	1	—	Initiator → Target
Status	—	1	1	Target → Initiator
Message-Out	1	1	—	Initiator → Target
Message-In	1	1	1	Target → Initiator

Table 2.3.
Overview of the various transmission phases and the drive signals MSG, C/D and I/O.

Transfer phases

Because they enable data or messages to be transferred via the datalines, the *Command, Data, Status* and *Message* Phases go under the header of *Information* or *Transfer* Phases.

A further feature shared by all Transfer Phases is that they are invariably controlled by the then active Target. Table 2.3 shows that each of the six different Transfer Phases may be set by '0' and '1' levels on the three control lines *MSG, C/D* and *I/O*. As explained in section 2.1, these control lines are exclusively reserved for bus access operations by Targets.

Message-Out Phase

A Message-Out Phase is used by a Target to send a message to the Initiator, or receive one from it. Depending on the transfer direction (seen from the Initiator), the term *Message-Out Phase* is used (after the Selection Phase, a message is sent by the Initiator to the Target), or *Message-In Phase* (after the Reselection Phase).

2.2 Bus phases

This way of conveying messages represents the simplest form of communication on the SCSI bus. As clarified by Fig. 2.3, the Message-In Phases are provided as fixed elements in the SCSI protocol. They always follow a selection phase and may be appended to any of the other transfer modes. However, to convey error reports and other unexpected events, they may also be started totally independent of current affairs on the bus (with the exception of the Bus-Free and Arbitration Phases). For example, if the Initiator wants to convey a message, it sets the ATN signal. The Target may then fetch the message. Although the Target may choose the actual instant for doing so, the message must be fetched after all command and data bytes have been transferred.

Depending on their application, the messages may have different lengths. The Message format may consist of a single byte, or be distributed across several bytes. More information on the different formats is presented further on when the individual processes on the SCSI bus are discussed in detail.

Message-In Phase

As implied by the terms *Message-In* and *Message-Out*, data transfer may take place bidirectionally. In other words, a Target may also send messages to an Initiator. Details of this type of operation are shown in Fig. 2.4.

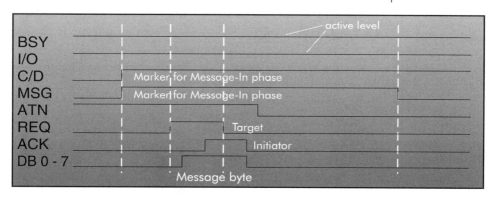

Figure 2.4.
Transmission of a Message byte from Target to Initiator via an REQ/ACK handshake.

The lines *BSY* and *I/O* are still active from the Reselection Phase. To mark the Message-In Phase, the Target also activates the signals *MSG* and *C/D* (see Table 2.3). It then puts the message byte on to the datalines and sets the *REQ* signal after a short delay (de-skew + cable skew delay). This is done to prompt the Initiator to receive the message. Reception is acknowledged by the Initiator with an *ACK* signal, whereupon the Target clears the message again and resets the *REQ* signal. As soon as the Initiator recognizes this procedure, it also resets the acknowledge signal *ACK*.

This handshake is referred to as *REQ/ACK Handshake* and is also frequently used in the other transfer phases. In case the message consists of several message bytes (the receiver may deduce pertinent information on this from the first byte), the previously outlined process is started again any number of times it takes to convey all message bytes. The Target then resets the *MSG* signal and so ends the Message-In Phase.

Command Phase

In accordance with Table 2.3, the Command Phase is marked by a set *C/D* signal, while the *MSG* and the *I/O* signal are at '0'. The *BSY* signal is at '1' in all transfer phases anyway, because the control of the bus was awarded to a SCSI device during the Selection Phase.

The Command Phase is employed for conveying single SCSI commands, which are described in greater detail in section 2.4. Here, we limit ourselves to a cursory look at the way the process evolves.

The SCSI commands are held ready by the Initiator, and must be processed by the Target. However, because the Target is currently in control of the bus (see above), it fetches commands from the Initiator using the REQ/ACK handshake sequence. The actual transfer is identical with that as described for the Message-In Phase. The Target sets the *REQ* signal, the Initiator puts the command byte on to the line and then sets *ACK* after a delay. The Target fetches the command byte and resets *REQ*. This is recognized by the Initiator, which responds by resetting *ACK*. The first command byte contains information concerning the number of bytes to be

2.2 Bus phases

conveyed, allowing the Target to find out how many times the cycle has to be repeated.

To close off the Command Phase, the Target clears the C/D signal.

Data-In (-Out) Phase

As indicated by the name, (user) data is conveyed in one direction or another during this phase. The I/O signal determines whether a *Data-In* or a *Data-Out Phase* is involved. As far as **data** transport is concerned, the Target and Initiator may agree upon the use of synchronous, and, consequently, faster data transmission. If one of the parties involved in the transmission (i.e., Target or Initiator) is unable to implement synchronous data transmission, then the data are sent over the bus is asynchronous mode, that is, using REQ/ACK handshaking as already described.

The asynchronous transmission consequently follows the usual pattern:

During the transmission from Initiator in the direction of the Target (Data Out, I/O=0), data are requested by the Target with

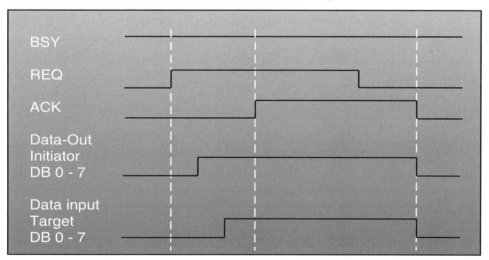

Figure 2.5.
Signal levels on the data and control lines during an asynchronous data transmission in the Data-Out phase.

the aid of the REQ signal. The data are put on the line by the Initiator and marked as ready for collecting with the aid of the ACK signal. The Target copies the data and resets REQ, whereupon the Initiator also clears ACK to '0'. Each additional cycle thus allows further data to be conveyed.

With data transport in the opposite direction (Data-In, I/O=1), everything happens just as described above, only the data are held ready by the Target using the REQ signal, while the Initiator uses the ACK signal to acknowledge correct reception. Next, the REQ signal is cleared by the Target, and the ACK signal is cleared by the Initiator. If necessary, a further cycle may then be initiated.

To clarify the difference between this process and synchronous data transmission, the transmission method using the REQ/ACK handshake in a Data-Out Phase is illustrated once more in Figure 2.5.

Synchronous data transmission
In principle, synchronous data transmission uses the same basic pattern as the asynchronous method. The only difference is that flexible instead of rigid definitions and agreements are applied. These 'internal' agreements between Initiator and Target are negotiated and arranged in the Message Phase. As already known, messages exchanged during this phase may have a length of several bytes.

The REQ/ACK handshake is also used in this case, only the rigid sequence *REQ signal–data transmission–ACK confirmation* is cancelled with synchronous data transmission. The purpose of this exercise is make the best possible use of the available buffers (organised as FIFO type shift registers) in all partaking devices, using a flexible arrangement. Time is saved in a simple manner when as many data sets may be conveyed, one after another, as can be contained in the receiver's buffer, without having to acknowledge receipt with an ACK after each individual data set. Acknowledging receipt is done in one go, that is, an equal number of ACK signals is transmitted as REQ signals were received before. The time saving is the result of smaller propagation losses on the line.

This is easily understood by a closer look at the events which occur during asynchronous transmission. The starting point is the set *REQ* signal supplied by the Target. This signal is required to travel over the line to be recognized by the opposite party (1st cable length). Next, the *ACK* signal is set by the Initiator. (2nd cable length), and recognized by the Target. Once the data transmission is finished, the Target returns the *REQ* signal to '0', which must be detected again by the other side (3rd cable length) to enable it to reset the *ACK* signal. The reset ACK signal must then travel to the Target again (4th cable length), enabling it, in turn, to set the *REQ* signal so that a new transmission cycle may commence.

That leaves us with the question how many *REQ* signals with appended data block may be transmitted in one go during a synchronous data transmission. The answer is supplied by the two devices agreeing upon a certain number during the Message Phase. In particular, the number of *REQ* signals plus associated data sets is defined that may be transmitted without causing an overflow of the receive buffer in the receiving device. In this way, the relevant devices negotiate the maximum number of cycles the *ACK* confirmation is allowed to trail the transmitted *REQ* signals. The term *REQ/ACK Offset* is used to refer to this value. Whenever the agreed REQ/ACK offset is reached, REQ signals may not be activated any more until the difference is decreased again (by received ACK signals) to a value which is smaller than the maximum agreed upon earlier. Figure 2.6 shows respective examples for synchronous data transmission during the Data-In and Data-Out Phases. The third sub-drawing illustrates the relations in the case of a defined REQ/ACK offset.

Fast-SCSI

Using this type of data transmission, the maximum data transfer rate of about 3.3 MByte/s with asynchronous transfer could be increased to 5 MByte/s with synchronous transfer, without any modifications. Synchronous data transfer was at least specified as an option under SCSI-1. Under SCSI-2, synchronous data transfer was enhanced further by shortening the maximum data settling times and the applied delays. The result was the name Fast-SCSI. Consequently, Fast-SCSI devices must be able to use synchronous data transfer, as well as handle the shortened timing values.

2. SCSI Basics

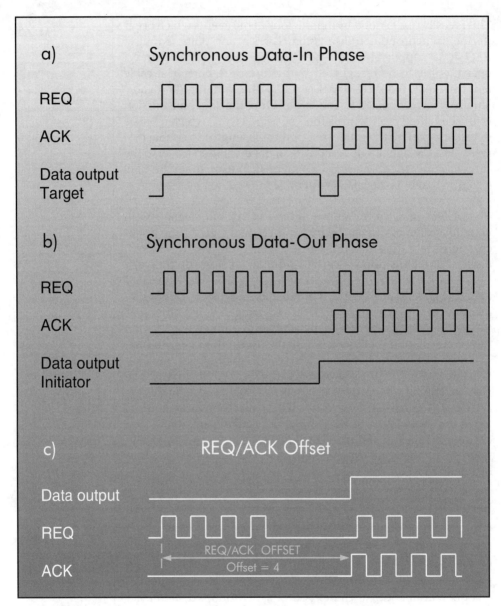

Figure 2.6.
These level diagrams clarify the synchronous data transmission in the Data-in (a) as well as the Data-out phase (b). The shift between REQ and ACK levels (REQ/ACK offset) is shown in part (c) of the illustration.

During the Message Phase, the devices involved in the data transfer agree upon the minimum timing and delay values, as well as on using synchronous data transmission. The two SCSI devices that wish to agree upon a data transfer face one another like two old school pals, who have been out of touch for a long time, and are now boasting their achievements: *my house, my car, my boat ...* . Only in this case, the boasting is about the timing values, which is actually useless to the device with the better specifications in this respect. The explanation is that those values are agreed upon that both devices are capable of handling safely. In other words, the values which are actually used are determined by the slowest of the two devices.

The shorter timing values agreed upon by the devices increase the maximum data transfer rate considerably. Moreover, the maximum allowable cable length was reduced to 3 metres under Fast-SCSI, resulting in a further reduction of propagation losses. In this way, a data transfer rate of up to 10 MByte/s may be achieved in synchronous transfer mode under Fast-SCSI. The highest achievable transfer rate on the eight-bit wide databus was thus trebled as compared with asynchronous transmission.

Wide-SCSI
Wide-SCSI is a promising extension for SCSI-2 which is sure to afford even higher data transfer rates to SCSI systems. This optional extension operates with 16 or even 32 datalines instead of eight (plus parity check). Consequently, a Wide-SCSI bus has a width of 16 or 32 bits. Because here, too, synchronous data transfer and the shortened SCSI timing may be negotiated between devices, the highest possible data transfer rate is doubled (16 bit) or quadrupled (32 bit). Under Wide-SCSI, data transmission rates of 10 Mbyte/s or even 40 MByte/s may be achieved if special Wide-SCSI cables are employed.

The development of the Message Phase shows a clear trend towards the function of a negotiation spot, because devices involved in a data transfer operation have to reach an agreement regarding the use of 8-bit, 16-bit or 32-bit wide datalines, in addition to synchronous data transmission and Fast-SCSI timing.

2. SCSI Basics

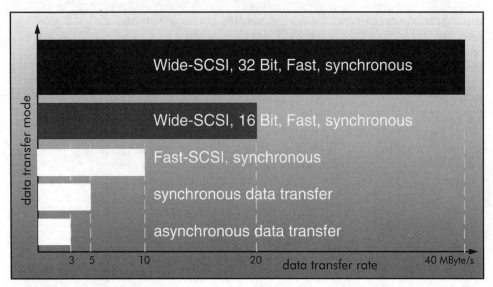

Figure 2.7.
Overview of achievable data transfer rates in the respective SCSI modes.

Figure 2.7 shows an overview of the highest data transfer rate that may be achieved in each of the available SCSI modes. This once again clarifies that Wide-SCSI is all about widening the bus, while Fast-SCSI is the mode with the faster timing, which may also be applied under Wide-SCSI.

Wide-SCSI offers an additional, very welcome and useful benefit to the user. As already mentioned, a regular 8-bit wide SCSI bus allows up to eight different devices to be accommodated (including the SCSI adapter). Each SCSI device has its own, individual, ID number (0 - 7), and is addressed by a logic '1' level on the corresponding dataline (DBn) during the Selection Phase. The limitation of only eight devices being separately addressable is caused by the fact that the bus has only eight datalines.

As explained earlier, Wide-SCSI offers 16 or even 32 datalines. Because the addressing of SCSI devices is also carried out with the aid of ID numbers, according to the same procedure, up to 16 (or 32) devices may be operated on a Wide-SCSI bus.

Exact information as regards the cable types that may be used, Wide-SCSI plugs and mixed use with devices designed with an 8-bit bus interface may be found in the chapter 'Practice' in this book.

Status Phase

The last of the transfer phases to be discussed is the Status Phase. As illustrated by Fig. 2.3, the Data Phase is always followed by a Status Phase. The Target uses a 1-byte long status byte to convey a status message to the Initiator. As shown in Table 2.3, the Target flags a Status Phase by setting the C/D and I/O lines to '1', while the MSG signal is deactivated, and, consequently, returned to '0'. The status byte is conveyed with the aid of a *REQ/ACK* handshake.

A Status Phase is always followed by a Message-In Phase. A completed command cycle is, for example, concluded with the message *Command Complete*, whereupon the bus is freed again (Bus-Free Phase).

Status messages not only occur at the end of a complete cycle, however, they are also inserted after aborted or rejected commands. Irrespective of the Phase to which a Status Phase is appended, the status message always originates from the Target, and is addressed to the Initiator.

After the description of all bus phases in the order shown in Fig. 2.3, the following section goes into greater detail. Not only the construction of individual bytes and byte sequences conveyed via the bus in the respective phases will keep us busy, but also the occurrence of errors and the way these are handled by the SCSI system.

2. 3 The SCSI Protocol

The individual phases that occur on the SCSI bus are known at this stage, as well as the order in which they follow one another within a 'normal' sequence. This section however discusses aspects like what should happen when, for instance, an Initiator wishes to abort an ongoing sequence, what is the function of a queue, and how is it implemented under SCSI-2, in which way is synchronous data transmission agreed upon during the Message Phase, and many more aspects.

Exception Conditions

Apart from the already familiar ten phases of the SCSI bus, which may be deemed the basic elements of data transmission under the SCSI protocol, there exist states which may be described as exceptions or transitions.

Attention Condition

This condition is marked by the activated ATN signal, and sets the Initiator. It may be compared with the raised hand of a pupil who wishes to communicate that he/she has something to say. During a transmission phase, the ATN signal must be activated no later than the end of the last ACK signal, or before the BSY signal is reset during the (re-)selection phase. The Target may respond at any time with a Message-Out phase, so collecting the message. A possibly fast response is useful, however, because nothing is known as yet about the meaning of such a message, so that acceptance should really not be delayed. The message should be fetched no later than after all commands and data bytes have been transmitted as a sequence.

To flag the end of the Attention Condition to the Target, the resetting of the ATN signal from the side of the Initiator must be accomplished before the last acknowledge signal (ACK) of the Message-Out phase.

Reset Condition

This condition is marked by the activated RST signal. For all practical purposes, the Reset Condition, and the RST signal triggered by it, is the emergency brake of the SCSI bus. After the RST signal has been set, all other signals must be reset to '0' after a predefined delay. Each transmission mode is ended immediately by this command. At the end of the Reset Condition (resetting of the RST signal), the bus goes into the Bus-Free phase.

2.3 The SCSI Protocol

Each SCSI device may trigger this condition at any time. As already mentioned, however, this is a kind of emergency brake which should not be used except in real emergency situations. Methods are available for regular error correction which proceed far more cautiously, and do not terminate the entire transmission just like that.

Nobody to talk to

The Selection and Reselection Phases serve to set up a connection between an Initiator and the desired Target. After the ID number has been set, a Target device has a maximum period of 200 ms left (*Selection Abort Time*) to fetch control of the bus by setting the BSY signal (see Section 2.2). What happens when a Target sees this period elapse?

To cope with this condition, a protection mechanism was built into the system which prevents the bus from freezing in this state. After the Initiator has withdrawn its own BSY signal, it allows the Selection Abort Time to elapse and end the Selection Phase, provided no Target has taken control of the bus. It is important to note how this is done by the Initiator.

At this instant, the ID numbers of Initiator and Target are set (via the respective datalines), while the SEL signal and the ATN signal (or the I/O signal in the Reselection phase) are available. Only after the signals on the datalines are cleared to '0', the control lines are withdrawn. If another sequence were used, a total blockage of the bus would occur in case the Target responds the instant the Initiator gives up.

This is only an example to illustrate that error conditions, which may arise in such an extensive procedure as the SCSI protocol, should be taken into account to prevent the odd (unwelcome) surprise.

2. SCSI Basics

Message codes

An extensive explanation was given earlier on about the fact that a Message phase serves to convey messages, whose direction is determined by the selection between a *Message-In* and a *Message-Out* phase. Also known at this stage is the fact that the length of a message may vary from one to several bytes.

The sub-section on the *Attention Condition* exemplified that these messages may also be sent outside a regular Command sequence. It was not yet mentioned, however, how the devices involved handle the different message lengths, how the various messages are identified, and which message types are available.

Message	Direction	Opcode	Implementation	
			Initiator	Target
Linked Command Complete	in	0Ah	O	O
Linked Command Complete with flag	in	0Bh	O	O
Message Parity Error	out	09h	m	m
Message Reject	in/out	07h	m	m
No Operation	out	08h	m	m
Release Recovery	out	10h	O	O
Restore Pointers	in	03h	O	O
Save Data Pointer	in	02h	O	O
Terminate I/O Process	out	11h	O	O

O = optional, m = mandatory

Table 2.4.
All one-byte long messages that may be employed in the Message phase.

2.3 The SCSI Protocol

Message	Direction	Opcode	Implementation Initiator	Target
Head of Queue Tag	out	21h	O	O
Ignore Wide Residue	in	23h	O	O
Ordered Queue Tag	out	22h	O	O
Simple Queue Tag	in/out	20h	O	O

O = optional, m = mandatory

Table 2.5.
All defined 2-byte messages.

Message	Direction	Opcode	Implementation Initiator	Target
Modify Data Pointer	in	00h	O	O
Synchronous Data Transfer Request	in/out	01h	O	O
Wide Data Transfer Request	in/out	03h	O	O
reserved (used under SCSI-1)		02h		
reserved		04h – 07h		
vendor specific		80h – FFh		

O = optional, m = mandatory

Table 2.6.
Extended messages.

During a Message phase the MSG and C/D signals are active ('1'). An active I/O signal indicates a Message-In phase. The first byte of a message contains information which say whether the message consists of one, two or more bytes (extended message).

2. SCSI Basics

The Message phase may not be interrupted if more message bytes are to be transmitted. Multiple messages (mind you, not messages bytes) may also be transmitted one after another in a Message phase, although it is possible that interruptions occur between individual messages.

Tables 2.4, 2.5. and 2.6 show all Messages with a length of one or two bytes, as well as the extended messages. The addresses marked as 'reserve' are intended for future use.

Bit \ Byte	7	6	5	4	3	2	1	0
0	Extended Message								
1	Length of Extended Message								
2	Extended Message Code								
3	Extended Message arguments								
n									

Opcode (Byte 2)	Message extension
00h	Modify Data Pointer
01h	Synchronous Data Transfer Request
02h	reserved (used under SCSI-1)
03h	Wide data Transfer Request
04h – 07h	reserved
80h – FFh	vendor-specific

Table 2.7.
Coding and format of the extended messages.

2.3 The SCSI Protocol

During the Message phase that follows on a Selection Phase, an Identify Message must be accomplished first. Also allowed are the Abort and the Bus Device Reset messages. Neither of the latter two does, however, allow a successful continuation of the command sequence. If any other type of message is to be conveyed, then the Target must initiate the Bus-Free phase.

The coding of extended messages, as well as their format, is given in Table 2.7.

At this point we continue with a three-part, alphabetical, list of the individual messages, along with short explanations. The first part contains all reports that may be issued by an Initiator. Part 2 is restricted to messages which emanate exclusively from a Target. Part 3, finally, lists reports which may travel either way across the link. Next, the meanings of individual terms not mentioned so far, for example, *queue*, *LUN*, *pointer*, and others, are explained in great detail.

Initiator messages

Abort

A cancelling report which can only be issued by an Initiator. It ends the connection with the addressed Target, which has to go into the Bus-Free phase. Connections and states of other Targets are not affected. If a queue was agreed upon with the relevant Target, it is terminated completely.

Abort Tag

This special cancel report enables the currently active process of a queue to be cancelled selectively without affecting the remaining operations. Next, the system is taken into the Bus-Free phase.

Bus Device Request

This report is used by the Initiator to trigger a hard reset in the addressed Target. As a matter of course, this report forces the Target to cancel all currently active operations. After a hard reset, the Target takes the system into the Bus-Free phase.

Clear Queue

A report which removes all I/O operations from an indicated queue. An Attention condition is generated, whereupon the system is taken into the Bus-Free phase.

Head of Queue Tag

This message causes a change in the queue order. An I/P operation is placed at the head of the queue. A currently active operation is not interrupted by this message.

Initiator Detected Error

A general fault report from the Initiator to the Target. The Target responds by repeating the last operation.

Message Parity Error

This report is issued if a parity error is detected during the message transmission. In addition, an ATN signal is set before the end of the ACK signal of the faulty byte. To be able to repeat the message, the Target must take the system back into a Message-In phase.

Ordered Queue Tag

This determines the processing order in a queue on the basis of the entry instant. A process which was forwarded earlier is completed before a process which arrived later.

Release Recovery

The Initiator informs the Target that it is allowed to cancel the ECA condition, which was set after an error in a logic unit (LUN) of the Target (see Initiate Recovery).

Terminate I/O Process

This report is used by the Initiator to press for a 'soft' termination of a currently active I/O process. A record which is about to be written may be finished to prevent data fragments on the magnetic medium (these are readily caused by a hard interruption using the Abort message).

2.3 The SCSI Protocol

Target messages

Command Complete

This message is used by the Target to explain to the Initiator that an I/O process was properly finished. After issuing a *Command Complete* report, the Target goes into the Bus-Free phase.

Ignore Wide Residue

This message is used by the Target to tell the Initiator which bytes are to be regarded as invalid at the end of a Wide-SCSI transmission. When a 16-bit wide databus is used, databits 16 through 31 and their parity bit are, in principle, declared invalid.

It may also happen, however, that the total number of bytes to be transmitted is not divisible by the transfer width (i.e., the number of datalines). In that case, a couple of filler bytes are transmitted which must be marked as invalid.

Linked Command Complete
Linked Command Complete with Flag

These reports are used instead of *Command Complete* within a series of several commands. The last command of this series is then closed off again with *Command Complete*.

Linked Command Complete with Flag is used in case the command (from a series) to be terminated has set a flag itself by means of its control byte.

Modify Data Pointer

With the aid of this report a Target is able to modify the value of the Data Pointer directly. A signed integer number is transmitted, which is added to the current value of the Pointer.

Restore Pointers

This report prompts the Initiator to replace the current pointer (command, data and status pointer) by the one which was stored last. The purpose of this process is, for example, to start a Data-Out phase again, because the Target discovered an error during the transmission.

Save Data Pointer

This report causes the Initiator to save the current value of the Data Pointer. Such a report is issued before each Bus-Free declaration. However, with long data transfers also, saving the Data Pointer may be useful to prevent a repeat of the whole transmission when an error occurs.

Messages for both directions

Disconnect

This message may be used by Target as well as Initiator to initiate the interruption of an existing connection.

When it wants to free the bus, the Target normally starts by transmitting a *Save Data Pointer Message*, whereupon it issues *Disconnect* (*Message-In phase*), and subsequently goes into the Bus-Free phase.

The Initiator may use the Disconnect report to prompt the Target to immediately free the bus (if necessary, with a previous save of the data pointer).

Identify

This message may be used by a device to establish the connection between Initiator and Target (or LUN and Target routine, see I/O Operations). The device is the one which has obtained control of the bus after the Arbitration phase.

Initiate Recovery

If an error has occurred in the Logic Unit (LUN) of a Target, this switches into the Extended Contingent Allegiance (ECA) Condition, and reports the condition with a status report and a subsequent Initiate Recovery. Not every Target is able to generate an ECA condition for error correction, and the Initiator has the right to reject this report.

Message Reject

This is the most generally used error correction in the exchange of messages during a Message phase. It is used, for example, if a device involved in the communication does not support a message which is only optionally implemented. A Message Reject is then returned to reject the message.

An Initiator may transmit *Message Reject* with a set ATN signal, at an instant which is not later than the end of the ACK signal of the last REQ/ACK handshake.

By contrast, a Target emits *Message Reject* by switching immediately (instantly after the faulty message) to the Message-In phase.

If the ATN signal is still active after the end of the Message reject report, the Target is automatically prompted to enter the Message-Out phase, and await subsequent messages.

Simple Queue Tag

When this report is received, Initiator or Target are free to insert an I/O process at any position in the queue. This allows, for example, a hard disk drive to optimize disk access under the control of its own (internal) logic.

Targets that want to use the Reselection phase to re-establish a broken connection with an Initiator **always** use this message to make for smooth insertions into ongoing operations.

Synchronous Data Transfer Request

This report is employed by SCSI devices involved in a transmission to agree upon synchronous data transfer. This should always be done after any restart of the system, or a SCSI reset.

A SCSI device which would like to use synchronous data transfer sends a *Synchronous Data Transfer Request* report to the other party in the communication. This message contains the desired transfer period and the size of the target REQ/ACK offset. The addressed SCSI device also responds with a Synchronous Data Transfer Request message, which, in this case, contains as parameters the transfer period and the offset size which the device is capable of handling (in as far as the values are worse than requested). If the selected device is powerful, the parameter requirements of the first device are confirmed.

However, if it is not capable of supporting synchronous data transfers, two answering options are available:

a) the request is rejected using a Message Reject, or

b) a Synchronous Data Transfer Request Message is given in reply, containing a value of nought for the REQ/ACK offset.

Both answers force asynchronous data exchange. The agreed transmission parameters need not be negotiated from scratch at the beginning of every transmission, but remain valid until the next SCSI or Bus-Free request.

Wide Data Transfer Request

Wide-SCSI, that is, data transport over 16 or 32 datalines, should also be agreed between devices involved in communication. This is accomplished with the aid of the *Wide Data Transfer Request Message* which evolves completely analogous to synchronous data transfer.

The device that wishes to employ wide-SCSI sends a Wide Data Transfer Request to the other party. This request has a marker (byte 3) which says whether 16-bit or 32-bit transfer width is desired. The addressed device also responds with a Wide Data Transfer Request message in which the desired transfer width is confirmed, or a smaller value is indicated.

If the device is not capable of partaking in a wide-SCSI transfer, two option are available for refusing wide-SCSI operation.

Either a transfer width of 0 is confirmed, or a message Reject report is sent out. The negotiated bus width normally remains accepted until the next SCSI reset or Bus Device reset.

Because it may happen that filler bytes are transmitted in the last transfer section, it may be necessary to append a *Ignore Wide Residue Message* (see above).

Structure of the SCSI Target

In the introductory chapter I explained that the combination of a data storage device (fixed or removable disk, CD-ROM, tape streamer, etc.) and a SCSI interface goes under the generic denominator 'SCSI device'. Such a SCSI device (device plus SCSI interface) is the usual form of a SCSI device which is operated on the bus, and, generally speaking, addressed via its SCSI ID (identification number). However, a Target is not restricted to the previously described physical shape. At least according to the SCSI standard a Target may consist of a SCSI interface and up to seven logic units (LUNs). These LUNs are then addressed at the same ID number, and only separated by the Identify process (see I/O Operations). A Target should have at least one logic unit (LUN 0). In practice, a

2.3 The SCSI Protocol

SCSI interface has at least one drive unit (which takes us back to the usual shape of a SCSI device).

In case several data carriers are bundled to form a single Target (rare in practice), these need not necessarily belong to the same device class. It is, therefore, possible for a single SCSI interface to operate in conjunction with a hard disk drive, a CD-ROM drive and a tape streamer, and appear on the bus as one Target containing three logic units. In practice, however, I have never come across such constructions – what comes nearest is a stack of hard disks or a set of printers which form a single Target.

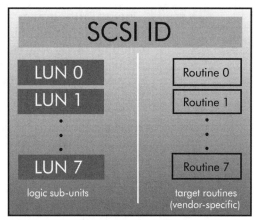

Figure 2.8.
Schematic structure of a SCSI Target.

In addition, so-called Target routines may be implemented in a Target. Such routines are vendor-specific processes which should be run in the Target, taking over diagnosis or maintenance tasks, and introduced for the first time under SCSI-2. The SCSI standard does not define requirements for these Target routines. They are just options for device manufacturers to accomplish further completion of the product, and are rarely used till now.

If available, a Target routine is addressed under the same SCSI ID, although a specific Identify message is used.

Tagged queues

The introduction of so-called tagged queues may be considered an important add-on to the SCSI protocol under SCSI-2. SCSI Targets which support *Tagged Queue* mode (the implementation of queues is optional) are capable of inserting executable commands into a queue, for processing in succession. A selection is available whether commands are processed in the order in which they arrived, or priorities may be given. This gives the Target the ability, for example, to optimise disk access.

Queues without tags already existed under SCSI-1, allowing Initiator commands to be executed in succession. Simultaneous communication with other Initiators was not possible. The introduction of tagged queues, that is, queues with an identification, enables Targets to accept commands for different I/O operations from different Initiators (up to 256), and process these commands in the order which is the most useful. The respective Initiator has available three messages, *Simple, Ordered and Head of Queue Tag Message*, to exercise control over the position of its commands in the queue.

Each of the above mentioned Queue Tag messages contains a unique reference to the respective I/O operation in the form of a number (the Queue Tag proper). Such a message is also transmitted by a Target after an interruption of an I/O operation, so that the Initiator knows, after the Reselection phase, which I/O operation is continued.

Rejections and error reports

As already mentioned, the implementation of tagged queues is optional. Consequently SCSI devices that do not support this mode must have the possibility to reject Queue Tag messages. This is done with the aid of a *Message Reject* report.

A possible source of errors in Tagged Queue mode (assuming the Target is capable of handling queues) is an incorrect queue identification. The insertion in the queue of a command under a wrong (i.e., already allocated) number by the Initiator (*Incorrect Initiator Connection*) causes the Target to respond with an interruption of all I/O operations on this LUN, and a subsequent Status Check.

By contrast, if a Target attempts the reconstruction of an I/O operation under a false Queue Tag (after a Reselection phase), this is punished by the Initiator with an *Abort Tag Message*.

I/O operations

You may find it astonishing at first glance that Latin terms have found their way into the otherwise very technical field of SCSI. After all, the ancient Romans never had computers or comparable technical achievements at their disposal. On the other hand, borrowing from an ancient language is not unusual when established, technically-related terms from one's own language are already in use, by definition, for other processes.

In SCSI terminology, the word *nexus* (Latin: connection, junction) describes the complete logic connection during an I/O process between an Initiator and a Target, including any sub-units involved. A special kind of syntax is employed to characterise different types of nexus. (It would have been interesting in this respect to hear a member of the X3T9.2 Committee of the ANSI Committee, who 'hatched' the SCSI standard, pronounce and emphasize the word *nexus*!)

The simplest connection option is an I_T nexus between an Initiator and a Target. If, ar the side of the Target, a distinction is made between different sub-units (LUNs), the term I_T_L nexus refers to the link between Initiator and Target LUN. Similarly, a junction between an Initiator and a Target Routine is identified as an I_T_R nexus. If it is not necessary to distinguish between LUN and Target Routine, the SCSI standard refers to an I_T_x nexus.

In contrast with a LUN, further extensions do not exist with a Target Routine. Each LUN of a Target is (optionally) able to work with a so-called *Tagged Queue*, a sorted row. Such a queue may hold up to 256 executable commands. A nexus between an Initiator and the Queue of a Target LUN is then, logically, called an I_T_L_Q nexus.
    In cases where it is irrelevant whether an I_T_x or an I_T_L_Q nexus is involved, the SCSI terminology uses the expression I_T_x_y nexus.

As a consequence of this excursion into Latin, the fact that there exist different 'targets' on a logic Target may create an assumption that there may also be different 'target addressing' types. To guarantee an error-free I/O process, that is, a data exchange between

2. SCSI Basics

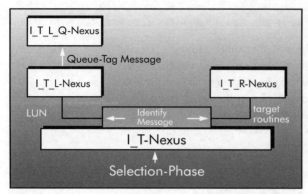

Figure 2.9.
Die unterschiedlichen Nexus-Varianten beim Aufbau eines I/O-Prozesses.

Initiator and Target sub-device, a differential message sequence is required to first establish the connection to the correct Target unit. Figure 2.9 shows such a target address, consisting of Selection phase, Identify phase and a concluding Queue Tag message. In this context, Queue Tag is the general header referring to the three messages *Head, Ordered* and *Simple Queue Tag Message.*

If the Initiator builds the connection (Selection phase), it addresses the desired Target (or its sub-device). In the Reselection phase, it is the Target that tells the Initiator, by means of the Identify message, which I_T_x nexus, and, thus, which I/O process, is to be continued.

Figure 2.10 shows the structure of an Identify message. A one-byte message is involved which, despite its brevity, may contain miscellaneous information.

Bit Byte	7	6	5	4	3	2	1	0
0	1	DiscPriv	LUNTAR	reserved		LUNTRN		

Figure 2.10.
Meaning of the individual bits in an Identify message.

Bit 7 is always at '1' and may be taken as the identification of this message.

2.3 The SCSI Protocol

DiscPriv: Disconnect Privilege
If this bit is set, the Target is granted permission to interrupt the currently active I/O operation, and declare the bus free. This bit can only be set by the Initiator.

LUNTAR: LUN/TARget routine
If this bit is at '1', a Target Routine is addressed. If it is at '0', the addressee is a LUN.

LUNTRN: the number of the LUN or Target Routine.

The installation of Target Routines in SCSI devices is optional, these routines being intended for diagnosis and maintenance purposes only.

Data Pointer

Because I/O operations may be stopped, interrupted and then continued again, the current conditions and data must be stored and given a reference. The Initiator offers storage areas for this purpose (in general, within the working memory of the host computer), and marks the current status reports, commands and data by Data Pointers. In this process, a distinction is made between active and saved Data Pointers. For each I/O range there exist three so-called Saved Data Pointers which identify the first byte in the data, command and status range of these I/O operations.

When the I/O operation is activated, the Initiator first copies the values of the Saved Data Pointers into the active pointer. In this way, the start state of the operation is created.

In the course of an I/O operation, it is possible for the Target to update the active Data Pointer values (*Modify Data Pointer*), or save them (*Save Data Pointer*) when an operation is interrupted. In this way, the I/O operation may continue at the point where it was interrupted, because the Initiator copies the saved pointers into the active ones at the start of the I/O operation.

When errors occur in a transmission phase, and they are recognized as such by the Target, this device may use the Restore Pointers Message to force the Initiator to return to the start of the transfer, and recommence the transmission.

Three of the four transmission phases (Command, Data and Status phase, see Figure 2.1) are monitored by this pointer system, and may be returned to their initial states, if necessary, or interrupted without problems. No pointers are available for the Message phase, however, which explains the fact that the completion of a message phase may not be interrupted (see *Message codes*).

2. 4 The SCSI commands

SCSI commands are conveyed to a Target during the *Command phase*. The Initiator uses a SCSI command to indicate what has to be done. The execution of the command is the Target's business. The logic connection between an Initiator and a Target is defined via a nexus, where it is entirely possible for several Initiators to stay in contact with a Target via different nexus junctions. This may result in advantages for the processing of individual commands, because each Initiator transmits commands or command sequences to a Target, which may decide on an order of execution which minimizes the complexity of the device.

An example: several Initiators want to access a single data carrier (hard disk, CD-ROM, etc.). In that case, the Target may fix the command execution order in such a way that adjacent ranges are read successively, even if the data are intended for different Initiators. This keeps head movement to a minimum and, consequently, reduces the read time for the total amount of data.

Command classes

A major advantage of the SCSI system is that different types of device are may be connected to the SCSI bus. The differences between devices may, however, be so large that it is necessary for SCSI commands to be adapted to a number of relevant device classes.

As a result, there are SCSI commands which are applicable to all device. classes, Together, these form a kind of main command set. There are, however, also commands which conform with peculiarities of a certain device class, and these may be used with devices of that class only.

2.4 The SCSI commands

Device classes

Before discussing the SCSI commands for individual device classes, it is first necessary to present the different *SCSI Device Types*.

Under SCSI-2, 10 different device classes are defined, and one general class. An overview is shown in Figure 2.11.

Within a class, all devices are treated equally.

All Device Typ	Generic device class
Direct-Access Devices	Disk drive units
Sequential-Access Devices	Tape drive units
Printer Devices	Printers
Processor Devices	Processors
Write-Once-Devices	WORM-/CD-R devices
CD-ROM Devices	CD-ROM drives
Scanner Devices	Scanners
Optical Memory Devices	Optical memories
Medium-Changer Devices	Medium changers
Communication Devices	Devices for data traffic

Figure 2.11.
SCSI device classes.

Each device class has its own command set and device-specific parameters. Likewise, commands and parameters are available for the general class that may be used on all device classes.

Mode parameters

In addition to its own command set, Each device class also has specific parameters which fix the device characteristics. These parameters are set and read with the aid of the *Mode Select* and *Mode Sense* commands respectively, and therefore go by the name of *Mode Parameters*.

The *Mode Parameters* of each device class are sub-divided into the *current*, *vendor* and *saved parameters*. The vendor parameters form part of the firmware, and are usually located in the EPROM on the SCSI interface inside the device.

The saved parameters are also located in a non-volatile memory. They may be changed if necessary, and are copied into the interface RAM when the device is switched on. There, they represent the *current* (up-to-date) values that are relevant for the operation of the equipment.

Command building

All SCSI commands contain at least the *Command Descriptor Block*, to which specific parameters may belong. While the *Command*

2. SCSI Basics

Descriptor Block is transmitted during the Command phase, the parameters are conveyed during the Data (In/Out) phase.

The execution of a command may involve several data transmissions, where messages may be exchanged between individual transmission sections. Only the transmission direction is governed from within the command: either the Data-In or Data-Out phase may be employed. A reversal is not allowed within a command. Each command is closed off with the Status phase, during which the Initiator receives a status report in the form of a status byte.

Command structure

A SCSI command may have length of 6, 10 or 12 bytes. The structure of such a command does not depend on the length in as far as the sequence is concerned. The first byte of each command contains the Operation Code, consisting of a 3-bit (MSB) information word about the command group, and five bits that contain the actual SCSI command. The three bits that form the command group allow $2^3 = 8$ different command groups. An overview of these is given in Table 2.8. Although the two groups marked *vendor-specific* may be equipped with specific commands by the equipment manufacturer, this is rarely done. The length of the SCSI command is assigned to each command group, so that a Target may deduce from the command group bits whether it is dealing with a 6-, 10- or 12-byte command.

The second byte of a SCSI command may be looked upon as a compatibility byte with SCSI-1. It contains the number of the logic unit LUN which is to be addressed by this command. Under the SCSI-2 standard, the LUN is always conveyed with the aid of an *Identify Message*, while under SCSI-1 the same is done in the second byte of the SCSI command.

Group	Opcode	Description
0	00h-1Fh	6-Byte commands
1	20h-3Fh	10-Byte commands
2	40h-5Fh	10-Byte commands
3	60h-7Fh	reserved
4	80h-9Fh	reserved
5	A0h-BFh	12-Byte commands
6	C0h-DFh	vendor-specific
7	E0h-FFh	vendor-specific

Table 2.8.
Overview of the eight command groups which are defined by the three higher-order bits of the opcode.

2.4 The SCSI commands

At the beginning of this book I mentioned the addressing of logic blocks was one of the greatest advantages that came with the introduction of the SCSI system. This method helps to ensure that future hardware developments may be integrated into a computer system without running into problems.

The five lower-order bits of Byte 1 and the subsequent Bytes 2 and 3 of a 6-byte long SCSI command address exactly these logic blocks of a storage medium (in as far as that is necessary with the relevant command). A 6-byte SCSI command consequently makes available $5 + 2 \times 8 = 21$ bits for the addressing of logic blocks. These 21 bits allow $2^{21} = 2{,}097{,}152$ logic blocks of 512 bytes each to be addressed. So, a single 6-byte SCSI command allows storage volumes of more than 1 Gigabytes to be handled. Consequently, storage media having a relatively large capacity must be addressed using 10 or 12-byte commands.

The venerable SASI standard from which the SCSI standard was developed, offered only 6-byte commands. Because these cause the above mentioned limitations as regards memory addressing, partly identical commands exist under SCSI in 6-byte and 10-byte versions.

In the 10 and 12-byte commands, the bytes with numbers 2 through 5 are responsible for the addressing of logic blocks. As a result, $4 \times 8 = 32$ bits are available for memory addressing with 'long' SCSI commands. So, a memory volume with a maximum size of $512 \times 2^{32} = 2.2 \times 10^{12}$ bytes (2.2 Terabytes) may be addressed under SCSI, leaving a 'certain' margin for future developments.

Depending on the device type to be addressed or the command used, it may not be necessary to address logic blocks. In those cases, device or command specific parameters may be transmitted in this range.

After the addressing/parameter range follows a byte which indicates the amount of data conveyed using this SCSI command.
   With a SCSI command that addresses logic blocks, this location indicates the number of addressed logic blocks.

2. SCSI Basics

By contrast, a parameter-relevant command provides the length of the parameter list. Depending on the application, the indications provided by this data length byte have to be interpreted in different ways. If logic blocks are addressed by the relevant command, then an entry 0 in this byte indicates that 256 blocks are conveyed. In a command that transfers parameters, however, the same value indicates that no data are being conveyed.

If the command does not convey data, or address logic blocks, the data length byte has no meaning.

Depending on the size of the SCSI command (6, 10 or 12 bytes), the number of data length bytes varies also. The basic structures of 6, 10 and 12-byte commands are shown in Figures 2.12, 2.13 and 2.14. A 6-byte command has one data length byte, a 10-byte command, two, and a 12-byte command may have four bytes of this type.

Figures 2.12, 2.13 and 2.14 also show that each SCSI command is terminated by a so-called *Control Byte*. This byte becomes meaningful only when SCSI commands are linked, else, it is a remnant of the old SASI and SCSI-1 days. The structure is shown in Figure 2.15. The two vendor-specific bits should guarantee downward compatibility with SCSI-1, in which many commands are still defined by the equipment manufacturers. It is precisely this

Bit / Byte	7	6	5	4	3	2	1	0
0	Operation Code								
1	LUN			(MSB)				
2									
3	Logic Block							(LSB)
4	Data length								
5	Control byte								

Figure 2.12.
Structure of a 6-byte SCSI command. Here, only 21 bits are available for parameter transmission or addressing of logic blocks.

2.4 The SCSI commands

Bit Byte	7	6	5	4	3	2	1	0
0	\multicolumn{8}{c	}{Operation Code}
1		LUN				reserved		
2	(MSB)							
3				Logic Block				
4								
5								(LSB)
6				reserved				
7	(MSB)			Data length				
8								(LSB)
9				Control byte				

Figure 2.13.
Structure of a 10-byte SCSI command.

Bit Byte	7	6	5	4	3	2	1	0
0				Operation Code				
1		LUN				reserved		
2	(MSB)							
3				Logic Block				
4								
5								(LSB)
6	(MSB)							
7				Data length				
8								
9								(LSB)
10				reserved				
11				Control byte				

Figure 2.14.
Structure of a 12-byte SCSI command.

71

2. SCSI Basics

pair of bits that used to cause errors because of poor compatibility. These bits are, therefore, best omitted. According to the SCSI specification, they should not be used any more. The middle four bits are reserved for future applications, while bits 0 and 1 (*Link* and *Flag*) may be employed with command linking (optionally, though).

Bit Byte	7	6	5	4	3	2	1	0
	vendor-specific		reserved				Flag	Link

Figure 2.15.
Organization of the Control byte, which is always the last byte in a SCSI command.

A set *Link* bit enables several SCSI commands to be linked (chained) in an I/O process. These linked commands may then be executed one after another without the bus being freed. This serves to prevent a target from attempting to insert a command for another I/O process as it aims to optimize its operations. Also, this method allows time to be saved on the *Bus Free*, *Arbitration* and *Selection Phase*.

The *Flag bit*, on the other hand, may only be '1' if the Link bit is set. A set Flag bit causes the message *Linked Command Complete with Flag* to be transmitted at the end of a (linked) SCSI command, instead of the usual *Linked Command Complete*, allowing the command to be marked in the command chain.

The Status report

The end of each SCSI command is marked by a *Status phase* (see Figure 2.3), during which the Initiator is informed that the transmission of the command has been completed properly or, if that is not the case, which causes may be indicated for a failure, and which measures are to be taken.

Table 2.9 provides an exhaustive overview of the reports, with the main ones highlighted. The report *Good* confirms an interference-free transmission. Similarly, the report *Busy* indicates that the Target device is currently occupied.

2.4 The SCSI commands

Status Byte Code	Status report	Description
00h	GOOD	Command executed successfully.
02h	CHECK CONDITION	Command could not be executed. Error cause to be checked with *Request-Sense* command.
04h	CONDITION MET	Indicates that a desired state was reached. This condition is confirmed with *Condition Met* if, for example, the data have been loaded in the buffer memory with a Pre-Fetch command.
08h	BUSY	Target is currently busy, and unable to accept the (correct) message.
10h	INTERMEDIATE	Used instead of the status report GOOD within a command sequence.
14h	INTERMEDIATE-CONDITION MET	Used instead of CONDITION MET within a command sequence.
18h	RESERVATION CONFLICT	LUN is currently reserved for another Initiator, try again later.
22h	COMMAND TERMINATED	Message issued after an I/O process has been halted.
28h	QUEUE FULL	Indicates that the queue is currently unable to accept further commands.

Table 2.9.
A list of all possible status reports.

The report *Check Condition* is returned as status information whenever an error has occurred. Using the SCSI command *Request Sense* (see below), the Initiator may request further indications regarding the cause of the fault.

Generic commands

A plethora of SCSI commands is available. They may apply to different device classes, and differ in respect of their implementation type. There are commands which must be available (*mandatory*), while others exist that may be added by the manufacturer (*optional*).

Still other commands may be added by the manufacturer if and when required (*vendor-specific*).

Presenting and discussing each and every command in great detail would add several hundred pages to this book. Moreover, the actual value of such information would be relatively low for you, the reader, because such extensive command explanations and listings are at best interesting to developers of SCSI devices who, arguably, will be keen on going for the direct sources straight away, that is, the full SCSI standard documents. None the less, you may want to do that, too. The Appendix at the end of this book, and the information on the CD-ROM contain sufficient references to the SCSI standards and their sources.

Because of this, we will limit ourselves to a discussion of the main SCSI commands. First we look at the commands that are mandatory for all ten device classes. These are the SCSI commands *Inquiry*, *Request Sense*, *Send Diagnostic* and *Test Unit Ready*.

Test Unit Ready (opcode 00h)

This command enables the Initiator to find out whether the LUN has access to the physical device, or its medium. In this way, the Initiator may be informed, for example, whether or not a disk has been inserted in a removable disk drive unit, but also, during the processor boot-up time, whether a hard disk has reached its nominal rotation speed and is ready for use.

As far as the command structure and its operation are concerned, the *Test Unit Ready* command is a little different from normal SCSI commands because the actual heart or core of a SCSI command, the addressing or parameter range, is not used at all. In other words, this commands does not transfer parameters, and does not read data bytes. The actual use of the *Test Unit Ready* command is found in the Status report. If the SCSI device is ready for use, the status report *Good* is returned. In all other cases, *Not Ready* or *Check Condition* is returned, depending on the actual condition of the device.

After each SCSI command which is terminated or not executed as requested, the Target holds a report ready which covers the error cause (in the form of so-called *Sense Data*), which may be requested by the Initiator with the aid of the *Request Sense* command. It

2.4 The SCSI commands

Bit\Byte	7	6	5	4	3	2	1	0
0	TEST UNIT READY							
1	LUN			reserved				
2	reserved							
3	reserved							
4								
5	Control byte							

Figure 2.16.
Command structure of the Test Unit Ready command. The adressing/parameter range (second part of byte 1 to byte 4) is not used here.

should be noted, however, that only the Sense Data for the most recent SCSI command are available with the Target. The *Request Sense* command should, therefore, be transmitted straight away by the Initiator to make sure that no other SCSI command can reach the LUN in the mean time.

Inquiry (12h)

The Inquiry command is used by an Initiator to request all relevant data on a Target, including device type, SCSI options supported, operating parameters and manufacturer identification (vendor ID). This is a command which uses the possibility of parameter

Bit\Byte	7	6	5	4	3	2	1	0
0	INQUIRY							
1	LUN			reserved				EVDP
2	Page code							
3	reserved							
4	Data length							

Figure 2.17.
Structure of the Inquiry command.

2. SCSI Basics

Bit / Byte	7	6	5	4	3	2	1	0
0	Peripheral Qualifier			Device Type				
1	RMB ||||||| Device-type modifier (SCSI-1)						
2	ISO Version			ECMA Version			ANSI Version		
3	AENC	TrmIOP	reserved		Data format			
4	additional data length							
5 – 6	reserved							
7	RelAdr	WBus32	WBus16	Sync	Link	reserv.	CmdQue	SftRe
8 – 15	Vendor							
16 – 31	Product identification							
32 – 35	Product Revision Level							
36 – 55	vendor-specific							
56 – 95	reserved							
96 – n	vendor-specific							

Figure 2.18.
Structure of the individual bytes and their content with the return of the Inquiry parameters.

Peripheral Qualifier: These three bits indicate whether a device may be connected to this LUN, or is already connected. In the device status, no statement is made about device readiness.

Peripheral Device Type: These five bits are used to assign a SCSI device class to a physical peripheral device represented by a LUN.

Table 2.10.
Meaning of the individual bits and their possible content with the return of the Inquiry parameters.

RMB (Removable): Indicates that a removable medium is involved.

Device-Type Modifier: Possibility for the vendor (manufacturer) to implement device modifications under SCSI-**1**.

ISO Version: Indicates whether a device supports a certain ISO version (IS-9316).

ECMA Version: The same, ECMA-111.

ANSI Version: These indications may vary between 0 and 2, 0 indicates a SCSI-1 apparatus, 1 a SCSI-1 device which is CCS compatible, and 2, a SCSI-2 device.

AENC (Asynchronous Event Notification Compatibility): This bit concerns apparatus of the Processor Device class only. If the bit is set, the relevant apparatus supports asynchronous messages, i.e., it may be addressed by a SEND command.

TrmIOP: Device capable of processing *Terminate I/O Process* message

Data format: 0 = SCSI-1; 1 = SCSI-1 with CCS; 2 = SCSI-2

Additional datalength: Number of additional information bytes

RelAdr: In case the device supports chained commands, relative addressing may be used. In this system, the distance to the current logic block is addressed rather than the number of the logic block.

Wbus32: 32-bit Wide-SCSI is supported.

Wbus16: 16-bit Wide-SCSI is supported.

Sync: Device supports synchronous data transfer.

Link: Chained commands are supported.

CmdQue: Tagged queues may be processed.

Table 2.10. (continuation)
Meaning of the individual bits and their possible content with the return of the Inquiry parameters.

2. SCSI Basics

> **SftRe (Soft Reset):** When this bit is set, the device responds to a reset condition by a soft reset.
>
> **Vendor Identification:** Name of the manufacturer in plain ASCII code.
>
> **Product Identification:** Name or type of product in plain ASCII code.
>
> **Product Revision Level:** Version or revision number in plain text.

Table 2.10. (continuation)
Meaning of the individual bits and their possible content with the return of the Inquiry parameters.

exchange in a typical way. Because of this, the individual command sections will be examined in greater detail.

The structure of the Inquiry command is illustrated in Figure 2.17. The meaning of the terms and abbreviations in this drawing is as follows.

EVDP: *Enable Vital Product Data.* This bit is used to decide whether the Standard Inquiry Data are interrogated, or an extended form which is defined via the *Page Code*. A set EVDP bit activates the page code, causing the (optional) *Vital Product Data* (VDP) to be returned. If it is at '0', the standard values apply.

Page code: this is activated by the EVDP bit, and causes the VDP page to be read.

Allocation Length (data length): this pre-sets the maximum number of Inquiry Data to be anticipated. If this byte is set to FF_h, 256 bytes are reserved, and all available Inquiry Data may be received. A smaller value limits the number of parameters to be returned. With all parameters of the bytes greater than or equal to ≥ 36, only manufacturer-specific information is involved.

2.4 The SCSI commands

The parameters of the Inquiry Data are also structured, and are transmitted according to a fixed scheme as shown in Figure 2.8. Table 2.10 allocates the possible parameters to the bits shown in Figure 2.18.

Request Sense (03$_h$)

This SCSI command is used by the Initiator to request Sense Data from the Target. Sense Data are available to a LUN whenever the previous SCSI command was terminated with the status report *Check Condition* or *Command Terminated*.

The Sense Data provide accurate indications to the Initiator as to why a command could not be executed as desired.

The *Request Sense Command* has a structure which is similar to the previously mentioned Inquiry command. In this case, too, a corresponding entry in the field reserved for *Allocation Length* enables the amount of Sense Data to be reduced, FF$_h$ representing the maximum value.

Error feedback

In this relation, the manner in which the error is reported back is much more interesting than the structure of this command.

Let's look at the *Sense Data Format* as shown in Figure 2.19. The diagram shows that the transmission of the error cause is done in cumulative stages: first, a coarse orientation, then increasingly accurate information.

A Target addressed with Request Sense first replies with the error code 70$_h$ or 71$_h$. (70$_h$ = the status report Check Condition was generated because of the current SCSI command; 71$_h$ = the *Check Condition* message was prompted by a previous command). If there is no more to report, the Target resets the four Sense bits to zero (0$_h$ = *No Sense*).

By contrast, if the Target wants to produce error messages it may generate $2^4 = 16$ different Sense keys from the four Sense key bits (see Table 2.11). These allow a coarse classification of the error cause to be produced (e.g., 2$_h$ = *Not Ready*).

2. SCSI Basics

Bit\Byte	7	6	5	4	3	2	1	0
0	Valid	Error code						
1	Segment Number							
2	FilMrk	EOM	ILI	Res		Sense-Key		
3 – 6	Information							
7	Additional Sense length							
8 – 11	Command-specific information							
12	Extended Sense Code (ASC)							
13	Extended Sense Code (ASCQ)							
14	Field Replaceable Unit Code (FRUC)							
15	SKSV	C/D	reserved		BPV	Bit position		
16	Sense-key specific data: Byte position							
17								
18 – n	Additional Sense bytes							

Figure 2.19.
Sense Data format.

Valid: 0 = Bytes 3-6 invalid (no SCSI-2 standard); 1 = Bytes 3-6 valid.

Error code: 70h = Error code concerns current command; 71h = error code concerns earlier command.

Segment number: Indication of defective segment with *Copy* and *Verify* commands.

FilMrk: Position at file mark (1) **EOM:** Position at data carrier end (1) **ILI:** Error log. block length

Sense-Key: Standard error report (see Table 2.12).

Information: If the Valid bit is set, additional device-specific information may be found here.

Additional Sense length: Number of bytes to follow; Value = total number - 7.

ASC; ASCQ: Additional error codes (see Table 2.14). **FRUC:** vendor-specific.

SKSV: Sense key specific bytes valid (1) **C/D:** data output error (0), command output error (1)

BPV: 0 = byte error; 1 = bit error.

Bit position: Pointer to faulty bit **Byte position:** Pointer to faulty byte.

Table 2.11.
Description of the individual fields of the Sense Data format.

2.4 The SCSI commands

If this coarse error description is not sufficient, the bytes ASC (Additional Sense Code) and ASCQ (Additional Sense Code Qualifier) are used to convey much more detailed information about the state of the connected device or devices. All Sense Codes defined under SCSI-2 are shown in Table 2.14.

If the detailed information is still insufficient, the manufacturer may define *Additional Sense Bytes* which express the possible complaints produced by the relevant product.

0h	No Sense
1h	Recoverd Error
2h	Not Ready
3h	Medium Error
4h	Hardware Error
5h	Illegal Request
6h	Unit Attention
7h	Data Protect
8h	Blank Check
9h	Vendor Specific
Ah	Copy Aborted
Bh	Aborted Command
Ch	Equal
Dh	Volume Overflow
Eh	Miscompare
Fh	Reserved

Table 2.12.
The 16 different Sense keys.

No Sense	No data available.
Recovered Error	Although the last command was correctly finished, the Target had to take error correction measures.
Not Ready	Not possible to access LUN.
Medium Error	An error was detected on the data carrier.
Hardware Error	Target has discovered a hardware error (during command execution or self-test).
Illegal Request	Unacceptable command (possibly errors in parameters).
Unit Attention	State change of LUN (e.g., medium exchanged).
Data Protect	Data access denied.
Blank Check	Encountered unexpected data or blank ranges.
Copy Aborted	*Copy* or *Verify* commands aborted.
Aborted Command	Command aborted by Target.
Equal	Successful data comparison with *Search Data*.
Volume Overflow	Insufficient storage space on medium.
Miscompare	Original and copied data do not match.

Table 2.13.
Explanations of the individual Sense keys.

2. SCSI Basics

		D - DIRECT ACCESS DEVICE
		T - SEQUENTIAL ACCESS DEVICE
		L - PRINTER DEVICE
		P - PROCESSOR DEVICE
		W - WRITE ONCE READ MULTIPLE DEVICE
		R - READ ONLY (CD-ROM) DEVICE
		S - SCANNER DEVICE
		O - OPTICAL MEMORY DEVICE
		M - MEDIA CHANGER DEVICE
		C - COMMUNICATION DEVICE

ASC	ASCQ	DTLPWRSOMC (Device classes)	Description
13h	00h	D W O	address mark not found for data field
12h	00h	D W O	address mark not found for id field
00h	11h	R	audio play operation in progress
00h	12h	R	audio play operation paused
00h	14h	R	audio play operation stopped due to error
00h	13h	R	audio play operation successfully completed
00h	04h	T S	beginning-of-partition/medium detected
14h	04h	T	block sequence error
30h	02h	DT WR O	cannot read medium - incompatible format
30h	01h	DT WR O	cannot read medium - unknown format
52h	00h	T	cartridge fault
3Fh	02h	DTLPWRSOMC	changed operating definition
11h	06h	WR O	circ unrecovered error
30h	03h	DT	cleaning cartridge installed
4Ah	00h	DTLPWRSOMC	command phase error
2Ch	00h	DTLPWRSOMC	command sequence error
2Fh	00h	DTLPWRSOMC	commands cleared by another initiator
2Bh	00h	DTLPWRSO C	copy cannot execute since host cannot disconnect
41h	00h	D	data path failure (should use 40 nn)
4Bh	00h	DTLPWRSOMC	data phase error
11h	07h	W O	data resynchronization error
16h	00h	D W O	data synchronization mark error
19h	00h	D O	defect list error
19h	03h	D O	defect list error in grown list
19h	02h	D O	defect list error in primary list
19h	01h	D O	defect list not available
1Ch	00h	D O	defect list not found
32h	01h	D W O	defect list update failure
40h	NNh	DTLPWRSOMC	diagnostic failure on component nn (80h-ffh)
63h	00h	R	end of user area encountered on this track

Table 2.14.

Extended Sense codes (ASC and ASCQ) defined under SCSI-2.

2.4 The SCSI commands

ASC ASCQ	DTLPWRSOMC	Description
00h 05h	T......S..	end-of-data detected
14h 03h	T.........	end-of-data not found
00h 02h	T......S..	end-of-partition/medium detected
51h 00h	T.....O...	erase failure
0Ah 00h	DTLPWRSOMC	error log overflow
11h 02h	DT.W.SO...	error too long to correct
03h 02h	T.........	excessive write errors
3Bh 07h	..L.......	failed to sense bottom-of-form
3Bh 06h	..L.......	failed to sense top-of-form
00h 01h	T.........	filemark detected
14h 02h	T.........	filemark or setmark not found
09h 02h	...WR.O...	focus servo failure
31h 01h	D.L...O...	format command failed
58h 00h	......O...	generation does not exist
1Ch 02h	D.....O...	grown defect list not found
00h 06h	DTLPWRSOMC	i/o process terminated
10h 00h	D..W..O...	id crc or ecc error
22h 00h	D.........	illegal function (should use 20 00, 24 00, or 26 00)
64h 00h	....R.....	illegal mode for this track
28h 01h	........M.	import or export element accessed
30h 00h	DT.WR.OM..	incompatible medium installed
11h 08h	T.........	incomplete block read
48h 00h	DTLPWRSOMC	initiator detected error message received
3Fh 03h	DTLPWRSOMC	inquiry data has changed
44h 00h	DTLPWRSOMC	internal target failure
3Dh 00h	DTLPWRSOMC	invalid bits in identify message
2Ch 02h	.......S..	invalid combination of windows specified
20h 00h	DTLPWRSOMC	invalid command operation code
21h 01h	........M.	invalid element address
24h 00h	DTLPWRSOMC	invalid field in cdb
26h 00h	DTLPWRSOMC	invalid field in parameter list
49h 00h	DTLPWRSOMC	invalid message error
11h 05h	...WR.O...	l-ec uncorrectable error
60h 00h	.......S..	lamp failure
5Bh 02h	DTLPWRSOM.	log counter at maximum
5Bh 00h	DTLPWRSOM.	log exception
5Bh 03h	DTLPWRSOM.	log list codes exhausted
2Ah 02h	DTL.WRSOMC	log parameters changed
21h 00h	DT.WR.OM..	logical block address out of range
08h 00h	DTL.WRSOMC	logical unit communication failure
08h 02h	DTL.WRSOMC	logical unit communication parity error
08h 01h	DTL.WRSOMC	logical unit communication time-out
05h 00h	DTLPWRSOMC	logical unit does not respond to selection

Table 2.14. (continuation)
Extended Sense codes (ASC and ASCQ) defined under SCSI-2.

2. SCSI Basics

ASC ASCQ	DTLPWRSOMC	Description
4Ch 00h	DTLPWRSOMC	logical unit failed self-configuration
3Eh 00h	DTLPWRSOMC	logical unit has not self-configured yet
04h 01h	DTLPWRSOMC	logical unit is in process of becoming ready
04h 00h	DTLPWRSOMC	logical unit not ready, cause not reportable
04h 04h	DTL O	logical unit not ready, format in progress
04h 02h	DTLPWRSOMC	logical unit not ready, initializing command required
04h 03h	DTLPWRSOMC	logical unit not ready, manual intervention required
25h 00h	DTLPWRSOMC	logical unit not supported
15h 01h	DTL WRSOM	mechanical positioning error
53h 00h	DTL WRSOM	media load or eject failed
3Bh 0Dh	M	medium destination element full
31h 00h	DT W O	medium format corrupted
3Ah 00h	DTL WRSOM	medium not present
53h 02h	DT WR OM	medium removal prevented
3Bh 0Eh	M	medium source element empty
43h 00h	DTLPWRSOMC	message error
3Fh 01h	DTLPWRSOMC	microcode has been changed
1Dh 00h	D W O	miscompare during verify operation
11h 0Ah	DT O	miscorrected error
2Ah 01h	DTL WRSOMC	mode parameters changed
07h 00h	DTL WRSOM	multiple peripheral devices selected
11h 03h	DT W SO	multiple read errors
00h 00h	DTLPWRSOMC	no additional sense information
00h 15h	R	no current audio status to return
32h 00h	D W O	no defect spare location available
11h 09h	T	no gap found
01h 00h	D W O	no index/sector signal
06h 00h	D WR OM	no reference position found
02h 00h	D WR OM	no seek complete
03h 01h	T	no write current
28h 00h	DTLPWRSOMC	not ready to ready transition, medium may have changed
5Ah 01h	DT WR OM	operator medium removal request
5Ah 00h	DTLPWRSOM	operator request or state change input (unspecified)
5Ah 03h	DT W O	operator selected write permit
5Ah 02h	DT W O	operator selected write protect
61h 02h	S	out of focus
4Eh 00h	DTLPWRSOMC	overlapped commands attempted
2Dh 00h	T	overwrite error on update in place
3Bh 05h	L	paper jam
1Ah 00h	DTLPWRSOMC	parameter list length error
26h 01h	DTLPWRSOMC	parameter not supported

Table 2.14. (continuation)

Extended Sense codes (ASC and ASCQ) defined under SCSI-2.

2.4 The SCSI commands

ASC	ASCQ	DTLPWRSOMC	Description
26h	02h	DTLPWRSOMC	parameter value invalid
2Ah	00h	DTL WRSOMC	parameters changed
03h	00h	DTL W SO	peripheral device write fault
50h	02h	T	position error related to timing
3Bh	0Ch	S	position past beginning of medium
3Bh	0Bh	S	position past end of medium
15h	02h	DT WR O	positioning error detected by read of medium
29h	00h	DTLPWRSOMC	power on, reset, or bus device reset occurred
42h	00h	D	power-on or self-test failure (should use 40 nn)
1Ch	01h	D O	primary defect list not found
40h	00h	D	ram failure (should use 40 nn)
15h	00h	DTL WRSOM	random positioning error
3Bh	0Ah	S	read past beginning of medium
3Bh	09h	S	read past end of medium
11h	01h	DT W SO	read retries exhausted
14h	01h	DT WR O	record not found
14h	00h	DTL WRSO	recorded entity not found
18h	02h	D WR O	recovered data - data auto-reallocated
18h	05h	D WR O	recovered data - recommend reassignment
18h	06h	D WR O	recovered data - recommend rewrite
17h	05h	D WR O	recovered data using previous sector id
18h	03h	R	recovered data with circ
18h	01h	D WR O	recovered data with error correction & retries applied
18h	00h	DT WR O	recovered data with error correction applied
18h	04h	R	recovered data with l-ec
17h	03h	DT WR O	recovered data with negative head offset
17h	00h	DT WRSO	recovered data with no error correction applied
17h	02h	DT WR O	recovered data with positive head offset
17h	01h	DT WRSO	recovered data with retries
17h	04h	WR O	recovered data with retries and/or circ applied
17h	06h	D W O	recovered data without ecc - data auto-reallocated
17h	07h	D W O	recovered data without ecc - recommend reassignment
17h	08h	D W O	recovered data without ecc - recommend rewrite
1Eh	00h	D W O	recovered id with ecc correction
3Bh	08h	T	reposition error
36h	00h	L	ribbon, ink, or toner failure
37h	00h	DTL WRSOMC	rounded parameter
5Ch	00h	D O	rpl status change
39h	00h	DTL WRSOMC	saving parameters not supported
62h	00h	S	scan head positioning error
47h	00h	DTLPWRSOMC	scsi parity error
54h	00h	P	scsi to host system interface failure
45h	00h	DTLPWRSOMC	select or reselect failure

Table 2.14. (continuation)
Extended Sense codes (ASC and ASCQ) defined under SCSI-2.

2. SCSI Basics

ASC ASCQ	DTLPWRSOMC	Description
3Bh 00h	TL	sequential positioning error
00h 03h	T	setmark detected
3Bh 04h	L	slew failure
09h 03h	WR O	spindle servo failure
5Ch 02h	D O	spindles not synchronized
5Ch 01h	D O	spindles synchronized
1Bh 00h	DTLPWRSOMC	synchronous data transfer error
55h 00h	P	system resource failure
33h 00h	T	tape length error
3Bh 03h	L	tape or electronic vertical forms unit not ready
3Bh 01h	T	tape position error at beginning-of-medium
3Bh 02h	T	tape position error at end-of-medium
3Fh 00h	DTLPWRSOMC	target operating conditions have changed
5Bh 01h	DTLPWRSOM	threshold condition met
26h 03h	DTLPWRSOMC	threshold parameters not supported
2Ch 01h	S	too many windows specified
09h 00h	DT WR O	track following error
09h 01h	WR O	tracking servo failure
61h 01h	S	unable to acquire video
57h 00h	R	unable to recover table-of-contents
53h 01h	T	unload tape failure
11h 00h	DT WRSO	unrecovered read error
11h 04h	D W O	unrecovered read error - auto reallocate failed
11h 0Bh	D W O	unrecovered read error - recommend reassignment
11h 0Ch	D W O	unrecovered read error - recommend rewrite the data
46h 00h	DTLPWRSOMC	unsuccessful soft reset
59h 00h	O	updated block read
61h 00h	S	video acquisition error
50h 00h	T	write append error
50h 01h	T	write append position error
0Ch 00h	T S	write error
0Ch 02h	D W O	write error - auto reallocation failed
0Ch 01h	D W O	write error recovered with auto reallocation
27h 00h	DT W O	write protected
80h XXh — FFh XX	}	Vendor-specific.
XXh 80h — XXh FFh	}	Vendor-specific QUALIFICATION OF STANDARD ASC. ALL CODES NOT SHOWN ARE RESERVED.

Table 2.14. (continuation)
Extended Sense codes (ASC and ASCQ) defined under SCSI-2.

2.4 The SCSI commands

Send diagnostic (1Dh)

The implementation of this command which is mandatory for all device classes prompts a Target to run a self-test. The result of such a test is reported to the Initiator in the form of status byte *Good* or *Check Condition*. In that case, the self-test byte (byte 1, see Figure 2.20) is set.

Bit Byte	7	6	5	4	3	2	1	0
0	SEND DIAGNOSTIC							
1	LUN			PF	reserved	ST	DevofL	UniofL
2	reserved							
3 + 4	Data length							
5	Control byte							

Figure 2.20.
Structure of the Send Diagnostic command.

None of the other options of this command are mandatory, i.e., they are all optional.

If the *Self Test bit* is at 0, a number of different diagnostic parameters may be requested for testing different device types.

The *PF bit* then determines the page format. If it is set, the page format complies with the SCSI-2 standard. If it is at '0', the page format is determined by the manufacturer.

DevofL means Device Offline. If this bit is set, a *Send Diagnostic* command may be used to get access to all LUNs and, if necessary, modify their states. An inactive *DevofL bit* prevents unwanted access, even when requested explicitly by the diagnostic parameters.

UnitofL stands for Unit Offline. This bit makes it possible to either allow all access to the addressed LUN (bit set), or disallow any state change (bit at '0').

The diagnostic parameters are defined for different device classes, and allow the use of manufacturer-specific test programs. The result of such a test may be requested on the Target with the aid of the *Receive Diagnostic Results* command.

Table 2.15 provides an overview of all SCSI commands, whether *mandatory*, *optional* or *vendor-specific*. It also shows which command is responsible for a certain device class. The four commands discussed on the previous pages are mandatory for all device classes and are, therefore, marked by an 'M' in all ten device class columns.

Six more commands also have a uniform definition for all device classes, albeit that they are optional as indicated by the 'O' mark in the device classes column. Specifically, these are the commands *Diagnostic Results* ($1C_h$), *Write Buffer* ($3b_h$), *Read Buffer* ($3C_h$), *Change Definition* (40_h), *Log Select* ($4C_h$) and *Log Sense* ($4D_h$).

The table also indicates which commands are mandatory for certain device classes, and which are only optionally implemented. Gaps in the device class columns indicate that the corresponding command is not defined for the relevant device class. This is self-evident because while it is understood that the command *Rewind* (01_h) makes sense for a tape streamer, it is useless for SCSI devices like hard disk drives or CD-ROM drives. The same goes, for example for the *Scan* ($1B_h$) command, which is indispensable for scanner devices, while causing raised eyebrows with devices of other classes.

The table is hexadecimally sorted. Commands appearing with different lengths are marked by a corresponding length entry in brackets after the command name.

Those of you who require more accurate information on the remaining commands are best referred to the contents of the CD-ROM supplied with this book. The CD-ROM allows you to retrieve all recommendations, commands and definitions of the SCSI-2 standard. Furthermore, both the CD-ROM and the following book sections mention a large number of www (world wide web) addresses which are used to disseminate the latest SCSI trends and innovations.

The table-oriented overview does not, however, close off the subject of SCSI commands. Even if I do not intend to go into further details regarding these commands, it is still necessary to elucidate the differences between the various device classes.

2.4 The SCSI commands

	D - DIRECT ACCESS DEVICE		*Column key*
	T - SEQUENTIAL ACCESS DEVICE		m = mandatory
	L - PRINTER DEVICE		o = optional
	P - PROCESSOR DEVICE		v = vendor-
	W - WRITE ONCE READ MULTIPLE DEVICE		specific
	R - READ ONLY (CD-ROM) DEVICE		
	S - SCANNER DEVICE		
	O - OPTICAL MEMORY DEVICE		
	M - MEDIA CHANGER DEVICE		
	C - COMMUNICATION DEVICE		

OP Code	D T L P W R S O M C (Device classes)	Description
00h	m m m m m m m m m m	TEST UNIT READY
01h	m	REWIND
01h	o v o o o o	REZERO UNIT
02h	v v v v v v	
03h	m m m m m m m m m m	REQUEST SENSE
04h	o	FORMAT
04h	m o	FORMAT UNIT
05h	v m v v v v	READ BLOCK LIMITS
06h	v v v v v v v	
07h	o	INITIALIZE ELEMENT STATUS
07h	o v v o o v	REASSIGN BLOCKS
08h	m	GET MESSAGE(6)
08h	o m v o o o v	READ(6)
08h	o	RECEIVE
09h	v v v v v v	
0Ah	m	PRINT
0Ah	m	SEND(6)
0Ah	m	SEND MESSAGE(6)
0Ah	o m o o v	WRITE (6)
0Bh	o o o o v	SEEK(6)
0Bh	o	SLEW AND PRINT
0Ch	v v v v v v v	
0Dh	v v v v v v v	
0Eh	v v v v v v v	
0Fh	v o v v v v v	READ REVERSE
10h	o o	SYNCHRONIZE BUFFER
10h	v m v v v	WRITE FILEMARKS
11h	v m v v v v	SPACE

Table 2.15.
All SCSI commands arranged in hexadecimal ascending order.

2. SCSI Basics

OP Code	D T L P W R S O M C	Description
12h	m m m m m m m m m	INQUIRY
13h	v o v v v	VERIFY(6)
14h	v o v v v	RECOVER BUFFERED DATA
15h	o m o o o o o o o	MODE SELECT(6)
16h	m m m m o	RESERVE
16h	m m m	RESERVEUNIT
17h	m m m m o	RELEASE
17h	m m m	RELEASE UNIT
18h	o o o o o o o o	COPY
19h	v m v v v	ERASE(6)
1Ah	o m o o o o o o o	MODE SENSE(6)
1Bh	o	LOAD UNLOAD
1Bh	o	SCAN
1Bh	o o o o	START STOP UNIT
1Bh	o	STOP PRINT
1Ch	o o o o o o o o o	RECEIVE DIAGNOSTIC RESULTS
1Dh	m m m m m m m m m	SEND DIAGNOSTIC
1Eh	o o o o o o	PREVENT ALLOW MEDIUM REMOVAL
1Fh		
20h	v v v v	
21h	v v v v	
22h	v v v v	
23h	v v v v	
24h	v v v m	SET WINDOW
25h	o	GET WINDOW
25h	m m	READ CAPACITY
25h	m	READ CD-ROM CAPACITY
26h	v v v	
27h	v v v	
28h	o	GET MESSAGE(10)
28h	m m m m m	READ(10)
29h	v v v o	READ GENERATION
2Ah	o	SEND MESSAGE(10)
2Ah	m m m	WRITE(10)
2Bh	o	LOCATE
2Bh	o	POSITION TO ELEMENT
2Bh	o o o o	SEEK(10)
2Ch	v o	ERASE(10)
2Dh	v o o	READ UPDATED BLOCK
2Eh	o o o	WRITE AND VERIFY(10)

Table 2.15. (continuation)
All SCSI commands arranged in hexadecimal ascending order.

2.4 The SCSI commands

OP Code	D	T	L	P	W	R	S	O	M	C	Description
2Fh	o					o	o		o		VERIFY(10)
30h	o					o	o		o		SEARCH DATA HIGH(10)
31h								o			OBJECT POSITION
31h	o					o	o		o		SEARCH DATA EQUAL(10)
32h	o					o	o		o		SEARCH DATA LOW(10)
33h	o					o	o		o		SET LIMITS(10)
34h								o			GET DATA BUFFER STATUS
34h	o					o	o		o		PRE-FETCH
34h		o									READ POSITION
35h	o					o	o		o		SYNCHRONIZE CACHE
36h	o					o	o		o		LOCK UNLOCK CACHE
37h	o								o		READ DEFECT DATA(10)
38h							o		o		MEDIUM SCAN
39h	o	o	o	o	o	o	o				COMPARE
3Ah	o	o	o	o	o	o	o				COPY AND VERIFY
3Bh	o	o	o	o	o	o	o	o	o	o	WRITE BUFFER
3Ch	o	o	o	o	o	o	o	o	o	o	READ BUFFER
3Dh						o		o			UPDATE BLOCK
3Eh	o					o	o		o		READ LONG
3Fh	o					o		o			WRITE LONG
40h	o	o	o	o	o	o	o	o	o	o	CHANGE DEFINITION
41h	o										WRITE SAME
42h						o					READ SUB-CHANNEL
43h						o					READ TOC
44h						o					READ HEADER
45h						o					PLAY AUDIO(10)
46h											
47h						o					PLAY AUDIO MSF
48h						o					PLAY AUDIO TRACK/INDEX
49h						o					PLAY TRACK RELATIVE(10)
4Ah											
4Bh						o					PAUSE/RESUME
4Ch	o	o	o	o	o	o	o	o	o	o	LOG SELECT
4Dh	o	o	o	o	o	o	o	o	o	o	LOG SENSE
4Eh											
4Fh											
50h											
51h											
52h											
53h											
54h											

Table 2.15. (continuation)
All SCSI commands arranged in hexadecimal ascending order.

OP Code	D	T	L	P	W	R	S	O	M	C	Description
55h	o	o	o		o	o	o	o	o	o		MODE SELECT(10)
56h											
57h											
58h											
59h											
5Ah	o	o	o		o	o	o	o	o	o		MODE SENSE(10)
5Bh											
5Ch											
5Dh											
5Eh											
5Fh											
A0h											
A1h											
A2h											
A3h											
A4h											
A5h										m		MOVE MEDIUM
A5h						o						PLAY AUDIO(12)
A6h										o		EXCHANGE MEDIUM
A7h											
A8h							o					GET MESSAGE(12)
A8h				o	o	o						READ(12)
A9h						o						PLAY TRACK RELATIVE(12)
AAh										o		SEND MESSAGE(12)
AAh				o	o							WRITE(12)
ABh											
ACh						o						ERASE(12)
ADh											
AEh				o	o							WRITE AND VERIFY(12)
AFh				o	o	o						VERIFY(12)
B0h				o	o	o						SEARCH DATA HIGH(12)
B1h				o	o	o						SEARCH DATA EQUAL(12)
B2h				o	o	o						SEARCH DATA LOW(12)
B3h				o	o	o						SET LIMITS(12)
B4h											
B5h										o		REQUEST VOLUME ELEMENT ADDRESS
B6h										o		SEND VOLUME TAG
B7h						o						READ DEFECT DATA(12)
B8h										o		READ ELEMENT STATUS

Table 2.15. (continuation)

All SCSI commands arranged in hexadecimal ascending order.

2.5 SCSI device classes

Direct Access Devices

The collective term *direct access devices* is used to unite storage media which allow direct access to logic blocks. Although this refers particularly to hard disk drives, the term also covers diskettes, magneto-optical (MO) drives and RAM disks.

A common feature shared by all *Direct Access Devices* is that data blocks are stored on them for later use. In this process, each data block is given a unique logic block address. An Initiator may read or write data blocks from/to such a storage medium by using respective READ or WRITE commands. Likewise, other commands, which also entail write or read operations, may be employed to change data on the medium, or make sure they are correct using the VERIFY command.

Because RAM disks are also *Direct Access Devices*, changes on the data carrier may be volatile, i.e., gone when the supply voltage is switched off, or non-volatile (hard disks, diskettes. etc.), all depending on the medium type.

Mapping
The relation between logic block addresses and physical memory blocks is referred to as *Mapping*. Mapping should be organized such that the transition from one to the next memory block does not incur unnecessary delays.

Because of this, logic blocks are assigned to successive memory ranges, in other words, *Linear Mapping* is applied.

For clarification, Figure 2.21 shows the schematic structure of a disk storage medium. The smallest physical memory unit is a sector. Several sectors make up a concentric track. Both the top side and the bottom side of a disk are divided into many tracks.

To enable the same track to be read at the top side and the bottom side of the disk, the system has to change from head 0 to head 1. If another track is to be read or written, the position of the read/write head has to be altered accordingly.

2. SCSI Basics

With reference to the scheme in Figure 2.21, linear mapping resulting in the smallest possible delays would look like this:

The sectors rotating under the read/write head are read one after another, corresponding to logic block addresses 0 through n. Next, the system changes from head 0 to head 1, providing access to the sectors with addresses $n+1$ through $2n$. Next, the heads have to be positioned over an adjacent track, and the reading of sectors under head 0 (block addresses $2n+1$ through $3n$) and head 1 (block addresses $3n+1$ through $4n$) starts again.

Under the SCSI standard, the size of a logic block may lie between 1 byte and 64 kBytes, and need not even be identical on the same medium. These liberties are not taken in practice, however, the operating system usually fixing the block size (DOS, for example, uses a block length of 512 bytes). The free definition of the block size once again emphasizes that SCSI storage media may be employed with a wide range of operating systems.

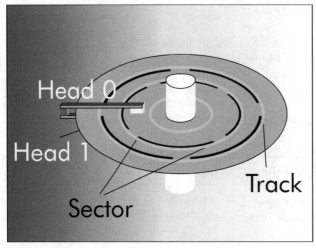

Figure 2.21.
Sectors, tracks, read/write heads – schematic structure of a Direct Access Device. A fixed disk may consist of several magnetic disks having the same construction.

LBN	Sector	Cylinder	Head
0	0	0	0
1	1	0	0
...	...	...	...
n	n	0	0
		Head change	
n+1	0	0	1
n+2	1	0	1
...	...	...	...
2n	n	0	1
		Track and head change	
2n+1	0	1	0
2n+2	1	1	0
...	...	...	...
3n	n	1	0
...			

Table 2.16.
Linear mapping of logic blocks.

2.5 SCSI device classes

In general, deviations from linear mapping may only be forced by defect memory ranges. Using this convention, faulty sectors are masked out by the SCSI interface of the relevant device, and the relevant logic blocks are allocated to other sectors. An operating system accessing such a SCSI device will not encounter defective memory ranges because these have been masked out beforehand by the intelligent controller.

The manner in which such memory defects are corrected is left to the manufacturer. After all, device-internal corrections are involved. In general, so-called defect lists are applied to the storage medium, containing bad-sector entries. A so-called *Plist - primary defect list* exists which has been drawn up by the manufacturer on checking the disk before it is shipped. The entries in the list indicate manufacturing defects. All other lists (*Glist - grown defect list,* etc.) contain defects which have been noted during formatting, or at a later stage. The Plist can not be altered. By contrast, all other lists may be updated as required.

Generally speaking, the user will not notice the occurrence of such errors. However, in case internal error correction is insufficient with a Direct Access Device, the Initiator still has the possibility to bring up a corrected mapping by using the (optional) *Reassign Blocks* command.

Direct Access Devices with removable data carriers (for example, Syquest drives, but also diskettes) do not make use of such a correction option, not because that would cause problems under SCSI, but because it must be possible to use such removable media on non-SCSI systems also, and these systems may not have an in-built fault correction mechanism like the one described.

Data cache

Most Direct Access Devices offer an integrated cache memory. Such a volatile memory area allows data to be stored temporarily with read as well as write operations, thereby improving the overall performance of the relevant device. An Initiator is not normally able to access a cache range directly.

With read operations, the cache memory is used for intermediate storage of data ranges of which it is likely that they have to be read as well (Pre-Fetch), and so reduce the access time. Here, too, linear mapping provides useful assistance. When the SCSI device is prompted to read a certain logic block, it is highly likely that subsequent blocks are also needed. Because of the linear relation between logic blocks and physical memory areas, the following sectors of a track are read also. The read head of the data memory is above this track anyway, so that reading an entire track does not take very much longer than reading the individual sector. If the relevant sectors are actually used (*Cache Hit*), the transmission time is considerably reduced because the data are already available in the cache memory, and no adjustments are required to find the relevant disk track.

If data have to be written to the storage medium, they are loaded into the write cache. The Target is then immediately able to transmit a status report to the Initiator, indicating that its write operation was successfully completed. This method, again, considerably reduces time spent on waiting for the requisite answer. The data are then fed to the cache memory under the control of the Target. This is referred to as a *Write-Behind* Cache. Although the use of the write cache produces the largest increase in performance, adopting intermediate data storage with write operations also increases the risk of data loss.

If an error occurs during intermediate storage of data to be read, a second disk access operation is performed, in other words, the risk of data loss does not exist. Not so with the write cache. If, for instance, the supply voltage disappears before data could be fed from the cache to the magnetic disk, these data are usually lost. If data are written directly to the disk, that is, without intermediate storage, most of today's hard disk drives are capable of finishing a write action started on a sector, even when the supply voltage disappears. So, no data is lost in that case when the supply voltage disappears during a write operation. If, however, the data is located in computer RAM, this type of extra security is of little use because data not yet saved, and resident in non-volatile memory, is lost when the supply power disappears.

SCSI offers the user the choice between a greater risk and increased performance as the higher-valued feature. The command *Synchronize Cache* may be used to force the Target to write the data set directly to the physical storage medium, thereby eliminating an additional risk.

A cache memory is normally organized as a shift register (FIFO). Consequently, data which was read first is also the first to be overwritten when the register is full. The Initiator, however, is, however, able to anchor accurately defined logic blocks in the cache memory by using the SCSI command *Lock-Unlock Cache*, thus preventing these blocks from being overwritten (*Lock Cache*), or make them available for overwriting again (*Unlock Cache*).

The Initiator may use the *Pre-Fetch* command to have logic blocks loaded into the cache memory to which (faster) access is required later. The *Pre-Fetch* command does not, however, initiate a data transmission via the SCSI bus.

Removable disks

Removable media also belong in the class of removable media. They include diskette drives, MO (magneto optical) disks and conventional removables such as Syquest disks. Forgetting about the absence of error correction I already mentioned, these media are treated, in principle, like normal, permanently installed, data memories under the SCSI conventions – with one difference.

As indicated by the name, removable media may be removed and exchanged. Consequently, the system must be able to check whether a data carrier is inserted, and read and write operations are possible, when such a device is addressed. The condition of the relevant device may be checked with the aid of the command *Test Unit Ready*. If a data carrier is inserted (i.e., the medium is *mounted*), the device is in the *Ready* state. If the state of an empty drive is checked (*unmounted*), the Target returns the status report Check Condition with the sense key *Not Ready*.

Depending on type and construction, an active drive has to be initialised with the command *Start Unit*.

When removable media with a cache memory are used, the command *Prevent medium Removal* may be used to prevent removal of the data carrier. In this way the system can make sure that all data have been copied from the cache memory on to the medium before the medium is exchanged. Of course this kind of exchange blocking works only when the relevant device has no disk removal mechanism, or when such a mechanism may be locked in an effective way.

The command *Allow Medium Removal* disables the electronic lock on the drive again.

SCSI commands for Direct Access Devices

Table 2.17 lists all SCSI commands defined for these device classes. Here, again, a distinction is made between optional and mandatory commands.

The command Format Unit is discussed in some detail here to once again clarify the relation between physical memory ranges and logic blocks (linear mapping) and the correction of defective sectors.

Bit Byte	7	6	5	4	3	2	1	0
0	FORMAT UNIT							
1		LUN		FmtDta	CmpLst	Defect list format		
2	Vendor specific							
3	(MSB)			Interleave				
4								(LSB)
5	Control byte							

Figure 2.22.
The Format Unit command.

The command *Format Unit* prompts a Target to format the storage medium of a LUN. This command may be rejected by the Target with the aid of the *Reservation Conflict* status report, provided another Initiator has denied access to the LUN.

In all other cases, formatting is carried out as follows:

- the medium is physically formatted, where each sector is written on to the medium, with its header, checksum and data range, in accordance with the indicated Mode parameters.

2.5 SCSI device classes

Command	Opcode	Type (mandatory / optional)
CHANGE DEFINITION	40h	o
COMPARE	39h	o
COPY	18h	o
COPY AND VERIFY	3Ah	o
FORMAT UNIT	04h	m
INQUIRY	12h	m
LOCK-UNLOCK CACHE	36h	o
LOG SELECT	4Ch	o
LOG SENSE	4Dh	o
MODE SELECT(6)	15h	o
MODE SELECT(10)	55h	o
MODE SENSE(6)	1Ah	o
MODE SENSE(10)	5Ah	o
PRE-FETCH	34h	o
PREVENT-ALLOW MEDIUM REMOVAL	1Eh	o
READ(6)	08h	m
READ(10)	28h	m
READ BUFFER	3Ch	o
READ CAPACITY	25h	m
READ DEFECT DATA	37h	o
READ LONG	3Eh	o
REASSIGN BLOCKS	07h	o
RECEIVE DIAGNOSTIC RESULTS	1Ch	o
RELEASE	17h	m
REQUEST SENSE	03h	m
RESERVE	16h	m
REZERO UNIT	01h	o
SEARCH DATA EQUAL	31h	o
SEARCH DATA HIGH	30h	o
SEARCH DATA LOW	32h	o
SEEK(6)	0Bh	o
SEEK(10)	2Bh	o
SEND DIAGNOSTIC	1Dh	m
SET LIMITS	33h	o
START STOP UNIT	1Bh	o
SYNCHRONIZE CACHE	35h	o
TEST UNIT READY	00h	m
VERIFY	2Fh	o
WRITE(6)	0Ah	o
WRITE(10)	2Ah	o
WRITE AND VERIFY	2Eh	o
WRITE BUFFER	3Bh	o
WRITE LONG	3Fh	o
WRITE SAME	41h	o

Table 2.17.
SCSI commands defined for Direct Access Devices.

2. SCSI Basics

- Next, the relation between the physical sectors and the logic blocks is set up (linear mapping).
- Next, sectors marked as defective are replaced based on the information supplied by the error lists, and the relevant logic blocks are allocated new, functional sectors.

The correction based on the error lists is controlled by the *FmtDta bit* contained in the *Format Unit* command. If this bit is at '0', no error lists are conveyed, if it is at '1', these lists are taken into account.

The bit marked *Cmplst* (complete list) indicates whether the error list transmitted by the Initiator is complete (Cmplst = 1), or just an addition to the existing error list (Cmplst = 0).

In both cases, the Target may write errors discovered during the formatting operation to be performed, into a *Dlist* (defect list), which is subsequently copied into the *Glist*.

Interleave factor

The field marked *Interleave* determines the interleave factor which is applied during the formatting. This setting allows logic blocks to be positioned so that the highest possible transfer rate may be achieved between Initiator and SCSI device. At an interleave factor of zero the Target copies its internal values. An interleave of 1 fixes an interleave factor of 1:1, which means that successive logic blocks are stored as a continuous sequence on the physical medium. All other values are manufacturer-specific.

The interleave factor is particularly interesting for storage media having a low rotation speed and a small buffer memory.

This is best explained by referring to a device with a very bad data transfer rate. Let's assume that a buffer memory is either not available at all, or so small that it can not hold a complete sector. In both cases, the system would have to wait one complete revolution of the storage medium, between the reading of a certain sector and a physically immediately following sector. This would be necessary because in the absence of sufficient intermediate storage the second sector may only be read when the transmission of the first sector is finished. If that does not happen in the short period between the end of first sector and the start of the second, the disk has to finish a complete turn before the transmission of the second sector may be started. Depending on the slowness (i.e., rotation speed) of the disk, this may result in a delay which returns all the time. This delay would obviously reduce the data transfer rate to be achieved.

In the above case it would be useful to refrain from storing successive logic blocks in successive sectors. Instead, logic blocks 1, 2, 3 and 4 could be stored in sectors 1, 3, 2 and 4. In this way, another sector is always in between successive blocks, leaving enough time for the transmission of the first block before the head is above the second.
If that is still not sufficient, it is, of course, possible to insert more than one sector between two blocks.

This formatting information is conveyed with the aid of bytes 3 and 4 of the *Format Unit* command.

Sequential Access Devices

The device class referred to as Sequential Access Devices includes SCSI magnetic tape recorders. I am not aware of other device types which would fit into this class.
With magnetic tape recorders you have to assume that the data carriers are exchangeable, that is, readable on other devices. This creates the need of a standardized recording method. In practice, however, many recording formats are currently around that are not compatible with one another. The recording method used in the end is determined by the tape drive unit, and meaningless for the actual use of the tape device under the SCSI standard. If the device is capable of handling several recording formats, these may be selected with the aid of the *Mode Select* command.

In the SCSI device class called Sequential Access Devices a distinction is made between *physical elements* and *logical elements*.
The physical elements group describes and defines the physical requirements and conditions needed to store and retrieve data on such a medium. This actually defines factors like the tape recording method, tape model, track position, bit density, encoding,

2. SCSI Basics

etc., in other words, it sets up the basics of data recording using a magnetic tape drive unit.

Within the range of logical elements, it is shown how SCSI achieves storage of data on this type of medium.

Physical elements

Despite different recording methods used on different tape drive units, the latter may be divided into three groups:

- devices using parallel track recording,
- devices using serpentine track recording,
- devices using helical-scan recording.

Figure 2.23.
Schematic representation of serpentine track recording.

Figure 2.24.
Schematic representation of parallel track recording.

Figure 2.25.
Schematic representation of helical-scan track recording.

The recording methods not only differ in respect of the track arrangements, but also whether or not multiple tracks may be written at the same time. The number of simultaneously written tracks is referred to as *track group* (TrkGrp).

In case not all tracks are written at the same time, and the tape is simply reversed at the end to fill the remaining free tracks, the recording method is called *serpentine* track recording (see *Figure 2.23*).

2.5 SCSI device classes

The simplest way of using this recording method is to record one track with the tape running in the forward direction, while the next track is written in the reverse direction, the next in the forward direction again, and so on.

This recording method is applied in particular with devices that operate according to the QIC standard. The method allows a simple read/write to ne used which is moved across a small distance on each tape reversal. The disadvantage of this method is the relatively low data rate.

Certain multi-track devices write all parallel tracks simultaneously. In these, there is only one track group, and no tape reversing facility is available. The method is generally referred to as parallel track recording (see *Figure 2.24*). Using parallel track recording, quite acceptable transfer rates of about 800 kByte/s may be achieved at an average tape speed.

Both recording methods mentioned so far write recordable data on longitudinal tracks. Not so with the so-called helical-scan recording method (see *Figure 2.25*), which uses a rotating magnetic head to write diagonal tracks on the tape. This technique is also commonly applied in video recorders.

Track groups are not defined in the helical-scan recording system, which is widely used in DAT recorders because of its inherent advantages when the tape speed is relatively low.

Partitions

A tape cassette (volume) may be divided into individual partitions, much like a hard disk as most of you will know. Each partition represents its own, small, storage medium. A partition has the following markers indicating the start (*BOP x, begin of partition*), the end (*EOP x, end of partition*), and a position before the end of the partition, *EW x* (early warning). The latter marker provides a timely warning before the end of the partition (or the tape end), enabling any data which may still be under way for writing into this partition.

If a tape is inserted into the tape mechanism (the device is mounted), it is rewound to the start of partition 0 (as a rule, the start of the tape range used for data recording).

2. SCSI Basics

If the command *Rewind* is given within partition x, the tape is rewound to the start of partition x.

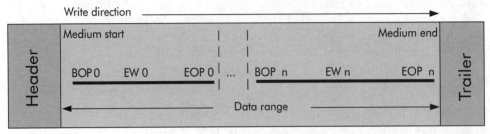

Figure 2.26.
Schematic representation of a medium divided into partitions having the required markers.

Any tape should have at least one partition (partition 0), while there is no upper limit. Figure 2.27 shows a feasible partition arrangement on one track each using serpentine recording.

In Figure 2.28, tracks are bundled pair-wise in one partition using the same recording method.

Figure 2.29 shows two partitions with the parallel-track recording methods. Here, the partitions correspond to a subdivision of the complete tape in individual sections.

Figure 2.27.
One partition per track group (serpentine recording).

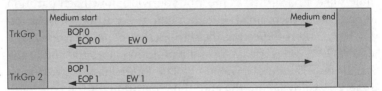

Figure 2.28.
One partition on two tracks groups (serpentine recording).

Figure 2.29.
Two partitions with the parallel track system (one track group).

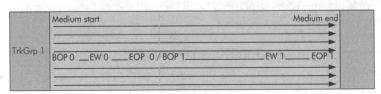

104

Three different options are available to create partitions on Sequential Access Devices:

- device-dependent using fixed location indicators,
- the Initiator indicates the number of partitions, and the arrangement on the medium is device-dependent,
- both the identification and the size of a partition are determined by the Initiator.

The indications as regards Initiator-controlled partitioning are conveyed via the *Mode Sense* and *Mode Select* commands.

Logic elements

Like a magnetic disk, a magnetic tape also presents itself to the Initiator as a sequence of logic blocks. At least two Initiator-accessible elements are contained between the *BOP x* and *EOP x* markers of a partition: the data blocks and the tape markers. These elements are controlled and transmitted using the commands *Read*, *Read Reverse*, *Write* and *Write Reverse*.

A data set which is written to tape by the Initiator, or read from tape, is referred to as a logic block. This unit corresponds to one or several physical blocks on a storage medium. If there is no 1:1 relationship between a logic block and physical blocks (as with linear mapping used on hard disks), the tape drive has to manage the individual fragments itself.

The maximum allowable length of a logic block is 16 Mbytes. It is, however, useful to write logic blocks having a smaller size, because a logic block has to be read or written without any interruption.

Tape marks

The Initiator uses tape marks to separate logic blocks and user data from identifiers. These marks are also used as jump marks that allow certain tape positions to be accessed without having to read all data beforehand. This is accomplished by the fact that tape marks are also recognized reliably when the tape is wound at high speed.

There are two types of tape mark: *File marks* and *Set marks*. The *Set marks* supply a coarse marker grid which is required for all tapes offering relatively high recording density, while *File marks* provide separations into smaller categories.

Block gaps

The space between individual physical blocks is called an *inter-block gap*. The smallest and largest size of such a gap are defined. On certain device types, the size of the inter-block gap may be defined by the Initiator with the aid of the *Mode Select* command. On most devices, however, the length cannot be changed.

Erase gaps also exist, these are written while an *erase* is performed on certain sections of the tape. Their smallest and largest lengths are also defined, although not all device types actually make use of these *erase gaps*. All devices using helical-scan recording, for example, omit them, and so achieve a somewhat higher net data density.

Depending on the recording standard, any exceeding of the maximum gap size is taken to indicate the tape end or a blank tape.

Data buffers

Two modes are defined for SCSI magnetic tape recorders: buffered and unbuffered. Consequently, tape drive units exist that do not have any kind of data buffer (a non-volatile intermediate memory), and can only be addressed in unbuffered mode.

The mode definition only refers to commands with write access, where it is immaterial whether data or tape marks are being written.

Analogous to the buffer memory of a hard disk, the buffered mode has the advantage that the data throughput with data buffer may be increased because no acknowledgement takes place before the data are physically written to the medium.

Switching between buffered and unbuffered mode (on devices having a data buffer) is possible using the *Mode Select* command.

All commands defined for this device class are listed in Table 2.18.

2.5 SCSI device classes

Command	Opcode	Type
CHANGE DEFINITION	40h	O
COMPARE	39h	O
COPY	18h	O
COPY AND VERIFY	3Ah	O
ERASE	19h	m
INQUIRY	12h	m
LOAD/UNLOAD	1Bh	O
LOCATE	2Bh	O
LOG SELECT	4Ch	O
LOG SENSE	4Dh	O
MODE SELECT(6)	15h	m
MODE SELECT(10)	55h	O
MODE SENSE(6)	1Ah	m
MODE SENSE(10)	5Ah	O
PREVENT/ALLOW MEDIUM REMOVAL	1Eh	O
READ	08h	m
READ BLOCK LIMITS	05h	m
READ BUFFER	3Ch	O
READ POSITION	34h	O
READ REVERSE	0Fh	O
RECEIVE DIAGNOSTIC RESULTS	1Ch	O
RECOVER BUFFERED DATA	14h	O
RELEASE UNIT	17h	m
REQUEST SENSE	03h	m
RESERVE UNIT	16h	m
REWIND	01h	m
SEND DIAGNOSTIC	1Dh	m
SPACE	11h	m
TEST UNIT READY	00h	m
VERIFY	13h	O
WRITE	0Ah	m
WRITE BUFFER	3Bh	O
WRITE FILEMARKS	10h	m

Table 2.18.
SCSI commands definded for the Sequential Access Devices class (O = optional, m = mandatory).

Printer devices

As will be seen in the practice-oriented section of this book, the *Printer Devices* class is an insignificant one. Printers driven via the SCSI bus (that is, the actual printer assembly, not a hard disk installed for storing fonts!) are few and far between, if found at all.

Command	Opcode	Type
CHANGE DEFINITION	40h	o
COMPARE	39h	o
COPY	18h	o
COPY AND VERIFY	3Ah	o
FORMAT	04h	o
INQUIRY	12h	m
LOG SELECT	4Ch	o
LOG SENSE	4Dh	o
MODE SELECT(6)	15h	o
MODE SELECT(10)	55h	o
MODE SENSE(6)	1Ah	o
MODE SENSE(10)	5Ah	o
PRINT	0Ah	m
READ BUFFER	3Ch	o
RECEIVE DIAGNOSTIC RESULTS	1Ch	o
RECOVER BUFFERED DATA	14h	o
RELEASE UNIT	17h	m
REQUEST SENSE	03h	m
RESERVE UNIT	16h	m
SEND DIAGNOSTIC	1Dh	m
SLEW AND PRINT	0Bh	o
STOP PRINT	1Bh	o
SYNCHRONIZE BUFFER	10h	o
TEST UNIT READY	00h	m
WRITE BUFFER	3Bh	o

Table 2.19.
SCSI commands for the Printer Devices class.

The SCSI device model for a printer is actually defined in basic terms only. The printer is then treated like a kind of black box which allows separate driving of the printer mechanism, paper feeding system (optional) and font management. No definitions exist on how this is to function in detail. Although a command set has been defined for this device class (see below), the response of a printer to the reception of a certain command is undefined. There are also parameter lists for the RS232 (serial) interface, but not for the Centronics interface which is far more widely used on printers.

It may be interesting to know that the SCSI printer model is built on a so called *bridge controller*, which means that the device interface is to drive different printers via different LUNs, although all printers share one SCSI ID. In this setup, the SCSI interface could either be housed in the printer enclosure, or it could be a stand-alone device which forms one SCSI device, being connected up to one or several printers.

The commands defined for the printer device class are shown in Table 2.19.

Processor devices

This device class has a truly multi-purpose nature and is really a way out for 'left-overs', or devices that do not fit in any of the other device classes. Although *SCSI processors* are 'only' capable of transmitting and receiving data, so that no standard device may be assigned to this class that may be purchased in any computer shop, this 'receiving' and transmitting' of data is, in itself, such a wide notion that Targets may be assigned to this group that would be difficult to fit in with another category.

An example of the use of this device class is the linking of two physically close computers (the limits as regards maximum cable length apply here, too). This SCSI model gives developers the opportunity to design suitable software and hardware which allows a SCSI-based connection to be achieved between PCs that monitoring and controlling measurement functions, and workstations that evaluate these results.

2. SCSI Basics

Admittedly, this may also be accomplished via an Ethernet link or similar, but remember that a possibly large application area should be covered when a SCSI device class is defined. As a result, device classes may get a relatively high or low significance. Only time will tell whether this device class will be used by standard devices. For one thing, the Processor Devices class is universally applicable and extendable.

Data exchange between *SCSI Processors* is accomplished in individual data packets. This is based on the assumption that both Initiator and target are aware of the manner in which data are exchanged, and how they are to be interpreted. No rules are available for this in the SCSI model.

The major transmission commands are *Send*, *Receive* and *Copy*.

A command overview is presented in Table 2.20.

Command	Opcode	Type
CHANGE DEFINITION	40h	O
COMPARE	39h	O
COPY	18h	O
COPY AND VERIFY	3Ah	O
INQUIRY	12h	m
LOG SELECT	4Ch	O
LOG SENSE	4Dh	O
READ BUFFER	3Ch	O
RECEIVE	08h	O
RECEIVE DIAGNOSTIC RESULTS	1Ch	O
REQUEST SENSE	03h	m
SEND	0Ah	m
SEND DIAGNOSTIC	1Dh	m
TEST UNIT READY	00h	m
WRITE BUFFER	3Bh	O

Table 2.20.
SCSI commands that may be used in the Processor Devices Class.

CD-ROM devices

Apparatus of this device type are far more widespread than those in the previously mentioned class. CD-ROM drives read data from a rotating medium — writing to a CD-ROM disc is not foreseen in this device class (see *Write Once Devices*). Data transmission may begin in any of the consecutively numbered data blocks, whose addressing is analogous to that applied with *Direct Access Devices*.

If a CD-ROM contains audio data which are to be transmitted via a separate audio output, the device itself has to arrange the control of these data, because these are not conveyed via the SCSI bus and can not, therefore, be affected directly by the usual SCSI command set. The same applies, in principle, to video data, in as far as they are conveyed via a separate output.

Data blocks on the CD-ROM

The audio CD is a kind of precursor of the CD-ROM. When the audio CD had been around for a number of years already, a standard was written for the recording of computer data, and laid down in the *Yellow Book*. The standards described in the *Yellow Book* also fix the way in which digital sounds are to be stored on a CD-ROM. These add-on standards make a distinction between two modes: Mode 1 for computer data, and Mode 2 for compressed music and/or video/graphic data.

The obvious question arises why new standards had to be created for the storage of computer data. After all, an audio CD also contains binary data, and as such there is no difference with computer data.

The underlying reason is the need of error correction. In the worst case, an audible noise is produced when an audio compact disc containing an uncorrected error is played on a CD player. If such an error would occur on a CD-ROM, for example, in a program code block, at least parts of the program would be unusable. That is why *Second Level Error Correction* (error detection/correction, EDC/ECC) has been defined in the *Yellow Book* for Mode-1 data.

2. SCSI Basics

Figure 2.30.
Block structure of the CD-ROM.

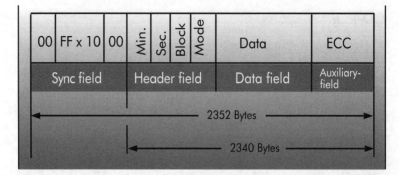

For this purpose, each track on a CD-ROM is divided into individually addressable sectors (*Frames*) of 2,352 bytes each. Each sector has a header which indicates the sector address and type.

As may be seen in Figure 2.30, a Mode-1 sector consists of a synchronization field with a length of 12 bytes, a header with a length of 4 bytes, user data with a maximum length of 2,048 bytes, and, finally, a 288-byte long *Auxiliary* field. The latter is subdivided into an error detection code with a length of 4 bytes, a reserved range of 8 bytes and an error correction range of 276 bytes. The error correction operates on the basis of a checksum comparison.

The sectors are counted in minutes, seconds and sector numbers starting from the inner track. 75 sectors form one second, 60 seconds form one minute, and up to 75 minutes fill a complete CD. The address format is designated *MSF* (minutes, seconds, frames).

A CD-ROM medium may be divided into a maximum of 99 tracks, where each *track* may only be occupied by sectors of the same type (Mode-1, Mode-2 or Audio). Tracks having a different sector type may not follow one another immediately; gaps, which have also been formatted, have to be inserted (Mixed-Mode CD-ROM). Addressing is carried out as already described by linear mapping of the physical sectors. In this system, audio tracks as well as formatted gaps are enclosed between tracks with a different sector type. To SCSI commands, however, neither the audio tracks nor the gaps are directly accessible. How these ranges are dealt with has to be arranged by the manufacturer (interface/driver) in accordance with the ISO-9660 standard. Optional audio commands then enable a SCSI CD-ROM drive to be operated like an audio CD player.

2.5 SCSI device classes

This makes the CD-ROM drive the only SCSI device on which not all logic blocks (tracks) of a medium may be accessed by all SCSI commands that may be used on this device. The *Read Command*, for example, is unable to access the transition areas or the audio tracks. The *Verify* command for CD-ROM drives contains options which allow a check to be run to see if such tracks are involved.

Audio tracks may be read using the optional audio commands. The classification *optional* has some restrictions here — either all audio commands have been integrated, or none at all. After all, a single command would not make sense because the complete command set for processing audio tracks should be available, if at all.

These days, manufacturers provide audio functions on practically all modern CD-ROM drives.

Command	Opcode	Type
CHANGE DEFINITION	40h	O
COMPARE	39h	O
COPY	18h	O
COPY AND VERIFY	3Ah	O
INQUIRY	12h	m
LOCK/UNLOCK CACHE	36h	O
LOG SELECT	4Ch	O
LOG SENSE	4Dh	O
MODE SELECT(6)	15h	O
MODE SELECT(10)	55h	O
MODE SENSE(6)	1Ah	O
MODE SENSE(10)	5Ah	O
PAUSE/RESUME	4Bh	O *
PLAY AUDIO(10)	45h	O *
PLAY AUDIO(12)	A5h	O *
PLAY AUDIO MSF	47h	O *
PLAY AUDIO TRACK/INDEX	48h	O *
PLAY TRACK RELATIVE(10)	49h	O
PLAY TRACK RELATIVE(12)	A9h	O

Table 2.21.
Commands that work on CD-ROM devices.

2. SCSI Basics

Command	Opcode	Type
PRE-FETCH	34h	o
PREVENT/ALLOW MEDIUM REMOVAL	1Eh	o
READ(6)	08h	o
READ(10)	28h	m
READ(12)	A8h	o
READ BUFFER	3Ch	o
READ CD-ROM CAPACITY	25h	m
READ HEADER	44h	o
READ LONG	3Eh	o
READ SUB-CHANNEL	42h	o
READ TOC	43h	o
RECEIVE DIAGNOSTIC RESULTS	1Ch	o
RELEASE	17h	m
REQUEST SENSE	03h	m
RESERVE	16h	m
REZERO UNIT	01h	o
SEARCH DATA EQUAL(10)	31h	o
SEARCH DATA EQUAL(12)	B1h	o
SEARCH DATA HIGH(10)	30h	o
SEARCH DATA HIGH(12)	B0h	o
SEARCH DATA LOW(10)	32h	o
SEARCH DATA LOW(12)	B2h	o
SEEK(6)	0Bh	o
SEEK(10)	2Bh	o
SEND DIAGNOSTIC	1Dh	m
SET LIMITS(10)	33h	o
SET LIMITS(12)	B3h	o
START STOP UNIT	1Bh	o
SYNCHRONIZE CACHE	35h	o
TEST UNIT READY	00h	m
VERIFY(10)	2Fh	o
VERIFY(12)	Afh	o
WRITE BUFFER	3Bh	o

Table 2.21.
Commands that work on CD-ROM devices (continuation).

Audio playback is started using the command *Play Audio*. At the same time, the Initiator may use this command to check if the addressed Target is capable of handling audio commands, or if the CD (-DA or -ROM) inserted in the drive contains audio tracks. If one condition is not satisfied, or both, the command is rejected with the aid of the status message *Check Condition*.

The SCSI commands which are valid for all CD-ROM devices are listed in Table 2.21. The asterisk with the marker 'optional' of the audio commands is a reminder that these commands may only be integrated completely, i.e., not individually.

Optical Memory devices

The device class Optical Memories is, in principle, capable of supporting different media which are optically readable. The following media classes are available:

- read only (CD-ROM);
- write-only (WORM; write once read multiple);
- erasable and re-writable.

Actually, the device classes CD-ROM and WORM (see below) form a kind of sub-class of the *Optical Memory Devices*. In these sub-classes, CD-ROM and WORM drives may be addressed in a more accurate manner, so that it is possible, though not useful, to control these drive types via the more general device class.

The possibility exists, however, that the two above mentioned, different, media may be inserted into a drive for optical storage. Consequently, an Initiator must use the *Mode Sense* command before each access operation to establish the media type involved.

Because such a check is performed before every access operation, it is possible to use media with different application ranges (for example, *read only* and *write once*). Although I have no indications about the existence of such a medium, the optical storage device class would allow the use of such a data carrier, in as far as it might become available in the future.

Devices capable of reading CD-ROMs as well as reading and writing other types of optical storage media are, however, already with us (for instance, the PD drive from Panasonic).

As regards their construction, *Optical Memory Devices* are similar to *Direct Access Devices* (magnetic disks). A basic assumption is, however, that storage media of considerable size and capacity are involved. Media access is therefore restricted to 12-byte commands, because these offer 32 bits for logic block addressing, thereby enabling the maximum storage volume of 2.2 Terabyte to be employed.

Interesting, because different from magnetic disk drives, is the treatment of defective blocks (*Defect Management*). Analogous to the *Reassign Blocks* command with Direct Access Devices, the command *Update Blocks* (optional) may be used to assign a new physical range to a logic block. The only difference is that with optical storage media access to the defective range is still possible (if necessary) using the command *Read Defect Data*.

Defect Management is however usually built on optional commands. Therefore, when drives from different manufacturers are involved, a check has to be run to establish the commands that are implemented in the end.

Table 2.22 lists all SCSI commands which are valid for this device class.

2.5 SCSI device classes

Command	Opcode	Type
CHANGE DEFINITION	40h	O
COMPARE	39h	O
COPY	18h	O
COPY AND VERIFY	3Ah	O
ERASE(10)	2Ch	O
ERASE(12)	ACh	O
FORMAT UNIT	04h	O
INQUIRY	12h	m
LOCK/UNLOCK CACHE	36h	O
LOG SELECT	4Ch	O
LOG SENSE	4Dh	O
MEDIUM SCAN	38h	O
MODE SELECT(6)	15h	O
MODE SELECT(10)	55h	O
MODE SENSE(6)	1Ah	O
MODE SENSE(10)	5Ah	O
PRE-FETCH	34h	O
PREVENT/ALLOW MEDIUM REMOVAL	1Eh	O
READ(6)	08h	O
READ(10)	28h	m
READ(12)	A8h	O
READ BUFFER	3Ch	O
READ CAPACITY	25h	m
READ DEFECT DATA(10)	37h	O
READ DEFECT DATA(12)	B7h	O
READ GENERATION	29h	O
READ LONG	3Eh	O
READ UPDATED BLOCK	2Dh	O
REASSIGN BLOCKS	07h	O
RECEIVE DIAGNOSTIC RESULTS	1Ch	O
RELEASE	17h	m
REQUEST SENSE	03h	m
RESERVE	16h	m
REZERO UNIT	01h	O
SEARCH DATA EQUAL(10)	31h	O

Table 2.22.
Valid SCSI commands for optical storage devices.

2. SCSI Basics

Command	Opcode	Type
SEARCH DATA EQUAL(12)	B1h	o
SEARCH DATA HIGH(10)	30h	o
SEARCH DATA HIGH(12)	B0h	o
SEARCH DATA LOW(12)	B2h	o
SEEK(6)	0Bh	o
SEEK(10)	2Bh	o
SEND DIAGNOSTIC	1Dh	m
SET LIMITS(10)	33h	o
SET LIMITS(12)	B3h	o
START STOP UNIT	1Bh	o
SYNCHRONIZE CACHE	35h	o
TEST UNIT READY	00h	m
UPDATE BLOCK	3Dh	o
VERIFY(10)	2Fh	o
VERIFY(12)	AFh	o
WRITE(6)	0Ah	o
WRITE(10)	2Ah	m
WRITE(12)	AAh	o
WRITE AND VERIFY(10)	2Eh	o
WRITE AND VERIFY(12)	AEh	o
WRITE BUFFER	3Bh	o
WRITE LONG	3Fh	o

Table 2.22.
Valid SCSI commands for optical storage devices (continuation).

Write-Once devices

As already mentioned, this device class is really a spin-off of the optical storage media class, so that only the differences are discussed briefly here.

A write-once device must be regarded as an archiving medium only. Once written, a logic block can not be modified anymore. Consequently, no *Update Block* commands are provided in this device class.

Because such a medium may be written to in several *Sessions*, as opposed to a single pass, the command *Medium Scan* has great significance because it allows the system to find out which blocks are blank, and which are already written. Before each read or write access operation, this command must be executed to enable ranges filled with data to be recognized.

The command *Format Unit* is not foreseen either with Write-Once devices, because the manufacturer supplies the media pre-formatted, allowing them to be mounted straight away.

Individual defective storage sectors are found more frequently with *Write-Once devices* than, for instance, with magnetic data carriers. As a result, the drives feature ingenious error correction mechanisms which differ between manufacturers, and allow data to be recovered. Depending on device type and manufacturer, different correcting levels may be selected also, where a minimum correction is always active.

Only when these device-internal correction systems no longer work, a check is requested within the SCSI command set via the status message *Check Condition*. In this way, additional correction measures may be initiated with the aid of suitable Sense data.

The SCSI commands for Write-Once devices are largely identical to those for Optical Memory devices. Only the following commands are not available because of the above mentioned reasons: *Erase, Format Unit, Update Block, Read Defect Data, Read Generation* and *Read Updated Block*.

2. SCSI Basics

Scanner devices

Scanners are devices that scan two-dimensional or three-dimensional objects, generating their digital image. Examples include a drawing, a text page, a photograph, or, in the medical field, a human organ.

For these devices, the SCSI system offers a transport mechanism that conveys collected data for further processing.

Basically, scanners may be divided into two type classes:

- a class in which the same operations and sequences are performed all the time;
- another class in which scan functions may be modified, and have to be set up before each scanner pass.

Depending on whether the scanned image is displayed as black and white, halftone or in colour, a different number of scanned values is in order, while the data format used to store these values may differ from case to case.

This fact is however meaningless for SCSI. The scanner generates the desired data and conveys them under the control of the commands it obtains from the Initiator. The Initiator and the scanner have to know how to handle the relevant data format.

Scanners generate the digital image in a two-dimensional perpendicular co-ordinate system (multiple-layer scans are performed with three-dimensional images) The y axis then indicates the scanning direction. The area to be scanned may be reduced or divided into multiple windows. The abscissa of such a scan window is always the top left-hand corner (Figure 2.31). The x-axis runs in positive direction from the left to the right. The maximum scan range extends from the co-ordinate origin 0,0 to the highest value for x and y.

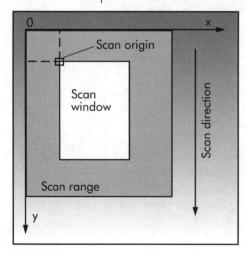

Figure 2.31.
Scanning co-ordinates.

2.5 SCSI device classes

The unit of measure used in the co-ordinate system (cm, inch, typographical point) is established via the *Mode Parameters*.

If you work with a scanner on which the scanning parameters have to be set before the actual scanning operation, quite a long series of parameters has to be transmitted to the scanner beforehand: origin of the scan window, unit of measure, window size, number of bits per scan value (pixel), contrast, brightness, and others. All these values are conveyed in the so-called *Window Descriptor*, which is illustrated in Figure 2.32.

If the Auto Bit in byte 2 is at 1, the scanner is able to generate sub-windows on its own within the defined scan window. This function is useful, for instance, if the scan range includes text as well as graphic elements, and the scanner (or the relevant software) is capable of treating text separately. The functions of bytes 1 to 28 are self-explanatory: they determine the resolution, scan origin, window size, brightness, the threshold for mono images, contrast, image composition, number of bits per pixel (gray steps or colour depth), and halftoning pattern.

The *RIF* bit in byte 29 indicates a positive or a negative image. The bytes that follow define the compression technique, or are reserved for manufacturer-specified functions.

All SCSI commands for scanners are listed in Table 2.23. It will be noted that the *Scan* command is only optionally implemented. This may seem odd, because one would assume that this com-

Bit / Byte	7	6	5	4	3	2	1	0
0				Window identification				Auto
1				reserved				
2 – 3				Resolution x-axis				
4 – 5				Resolution y-axis				
6 – 9				Origin x-axis				
10–13				Origin y-axis				
14–17				Window with				
18–21				Window length				
22				Brightness				
23				Threshold				
24				Contrast				
25				Image composition				
26				Bits per pixel				
27–28				Halftone pattern				
29	RIF			reserved				Filler
30–31				Bit order				
32				Compression method				
33				Compression parameters				
34–39				reserved				
40 – n				Vendor-specific				

Figure 2.32.
Structure of a Window Descriptor.

2. SCSI Basics

mand in particular is certain to be used. The explanation is quite simple. On most scanners, the actual scanning operation is started manually, and that obviates the need for a SCSI command.

Command	Opcode	Type
CHANGE DEFINITION	40h	o
COMPARE	39h	o
COPY	18h	o
COPY AND VERIFY	3Ah	o
GET DATA BUFFER STATUS	34h	o
GET WINDOW	25h	o
INQUIRY	12h	m
LOG SELECT	4Ch	o
LOG SENSE	4Dh	o
MODE SELECT(6)	15h	o
MODE SELECT(10)	55h	o
MODE SENSE(6)	1Ah	o
MODE SENSE(10)	5Ah	o
OBJECT POSITION	31h	o
READ	28h	m
READ BUFFER	3Ch	o
RECEIVE DIAGNOSTIC RESULTS	1Ch	o
RELEASE UNIT	17h	m
REQUEST SENSE	03h	m
RESERVE UNIT	16h	m
SCAN	1Bh	o
SET WINDOW	24h	m
SEND	2Ah	o
SEND DIAGNOSTIC	1Dh	m
TEST UNIT READY	00h	m
WRITE BUFFER	3Bh	o

Table 2.23.
Complete command set for SCSI scanner devices.

2. 6 Media-Changer devices

A SCSI media changer is a mechanical assembly that allows data carriers to be loaded from a 'stock' into a drive unit.

A media changer consists of the following four addressable elements:

- Medium Transport Element;
- Storage Element;
- Import Export Element (a mechanical unit, from which a medium may be removed by the user, or inserted);
- Data Transfer Element.

Although it may only contain a medium for a short period, each of these elements still has to be individually addressable. Any number of each element may be present in the equipment. At least with the Storage Element, that will normally be the case because this element is, by definition, capable of holding one medium only.

The individual elements of a medium changer may be addressed either via different SCSI IDs (unusual), or via a single ID but different LUNs.

This is easy to appreciate in the case of the *Data Transfer Element*, because that is a drive from a different SCSI class which requires its own addressing (LUN) in any case.

Consequently, the complete equipment, for example, a CD-ROM changer, belongs in two classes: the drive unit, in this case, goes to the *CD-ROM Devices*, while the changer proper goes to the *Medium-Changer Devices*.

An overview of the commands defined for SCSI Medium Changers is provided by Table 2.24.

Command	Opcode	Type
CHANGE DEFINITION	40h	O
EXCHANGE MEDIUM	A6h	O
INITIALIZE ELEMENT STATUS	07h	O
INQUIRY	12h	m
LOG SELECT	4Ch	O
LOG SENSE	4Dh	O
MODE SELECT(6)	15h	O
MODE SELECT(10)	55h	O
MODE SENSE(6)	1Ah	O
MODE SENSE(10)	5Ah	O
MOVE MEDIUM	A5h	m
POSITION TO ELEMENT	2Bh	O
PREVENT/ALLOW MEDIUM REMOVAL	1Eh	O
READ BUFFER	3Ch	O
READ ELEMENT STATUS	B8h	O
RECEIVE DIAGNOSTIC RESULTS	1Ch	O
RELEASE	17h	O
REQUEST VOLUME ELEMENT ADDRESS	B5h	O
REQUEST SENSE	03h	m
RESERVE	16h	O
REZERO UNIT	01h	O
SEND DIAGNOSTIC	1Dh	m
SEND VOLUME TAG	B6h	O
TEST UNIT READY	00h	m

Table 2.24.
SCSI commands for medium-changers.

Communication devices

The SCSI model for data communication devices defines the way in which data are exchanged between communication systems, via an electrical or a fibreglass cable, and using a unified protocol. The protocol is then not part of the SCSI model.

2.6 Media-Changer devices

This device class is similar to that of the Processor Devices, albeit that an additional addressing level is available in the data transfer model that allows different end devices to be addressed at the end of the transmission path.

The device class of *Communication Devices* enables the SCSI bus to be switched on to a network, and feed data packets into the network, or receive them from it.

The commands defined for this device class may be found in Table 2.25.

Command	Opcode	Type
CHANGE DEFINITION	40h	O
GET MESSAGE(6)	08h	O
GET MESSAGE(10)	28h	O
GET MESSAGE(12)	A8h	O
INQUIRY	12h	m
LOG SELECT	4Ch	O
LOG SENSE	4Dh	O
MODE SELECT(6)	15h	O
MODE SELECT(10)	55h	O
MODE SENSE(6)	1Ah	O
MODE SENSE(10)	5Ah	O
READ BUFFER	3Ch	O
RECEIVE DIAGNOSTIC RESULTS	1Ch	O
REQUEST SENSE	03h	m
SEND DIAGNOSTIC	1Dh	m
SEND MESSAGE(6)	0Ah	m
SEND MESSAGE(10)	2Ah	O
SEND MESSAGE(12)	AAh	O
TEST UNIT READY	00h	m
WRITE BUFFER	3Bh	O

Table 2.25.
SCSI commands for communication devices.

Small **C**omputer **S**ystem **I**nterface

Practice

Configuration
Cabels
Software Interface
SCSI BIOS
Adapters
Plug & Play
Devices

3. SCSI in practice

3.1 Configuration of the SCSI bus

A SCSI adapter, the link between SCSI bus and host computer, usually comes in one of two shapes — as an extension plug-in card, or as an on-board element on the PC's motherboard. Neither of the two versions has distinct advantages over the 'competition'. True, the use of an on-board adapter leaves one more slot available for additional extensions. On the other hand, a SCSI adapter in the form of an extension card offers you, the user, the freedom of choosing the actual adapter type (Wide-SCSI or others).

Generally speaking, a SCSI adapter has two connections for the SCSI bus — an internal and an external one. The internal link is used to connect all SCSI devices inside the computer case to the SCSI adapter, while the external connection provides the link to all SCSI devices outside the computer.

Figure 3.1 shows a SCSI adapter from Adaptec which comes as an insertion card. The black connector near the top of the picture

Figure 3.1.
Example of a SCSI adapter using the plug-in card construction (PCI version) – the external SCSI connection is at the left-hand side, the internal one, at the top.

3. SCSI in practice

is the internal bus connection. The external bus connector is fitted in the card mounting bracket, after all, it has to be accessible from the outside.

The electrical connection to the host computer is established by inserting the edge connector at the lower side of the adapter board into a free expansion slot on the motherboard. A drawing of the AHA 2920 which shows the above mentioned connectors is shown in Figure 3.2.

*Figure 3.2.
Sketch of the AHA 2920 and its connections.*

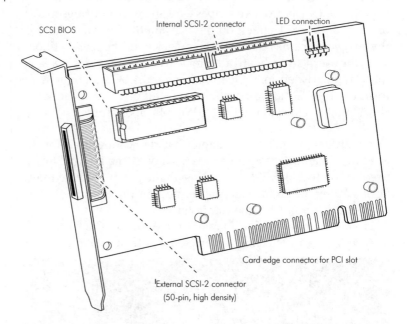

PCI or what?

You do not need a lot of imagination to figure out that the expansion slot into which the SCSI adapter is plugged has to be pin-compatible with the contact row on the edge of the adapter board, so that communication may be established with the host computer. There is, however, no such thing as a standard computer bus. Today, the bus types with the widest use on Windows/Intel systems are PCI and ISA. On server systems, the EISA bus is widespread, while the Microchannel and VL-bus are occasionally found in various computer systems. On Apple PCs, the PCI bus slowly supersedes older NuBus systems. Under UNIX and Amiga, ...

What was that again about SCSI being a universally applicable system?

Well, it is. SCSI adapters are available for each of the bus systems mentioned. Although the choice is of course dependent on the popularity of the desired computer bus, any system can be served. The largest choice of SCSI adapter models is currently available for the PCI and ISA buses. In this area, suppliers outclass one another all the time with new models, some of which may differ considerably as regards price and performance.

Obviously, the pinout of the bus connector strip is not the only difference between SCSI adapters for different bus systems. In particular the I/O processor on the insertion card must be able to cope with SCSI as well as the specifications of the desired computer system. The timing protocol, the voltage levels and many more aspects have to be matched to the relevant bus system. Consequently, a SCSI adapter for a less usual bus system may not be cloned just like that — it will have to be engineered largely as a new product.

In this way, price differences may be explained between adapters having similar performance, though aimed at different bus systems — it's the sales volume that determines the price.

If you have a choice between two bus systems (PCI and ISA), as on most Windows/Intel PC systems, the question arises for which bus system to purchase the SCSI adapter card. In that case, you should not only look at the price, but also at the achievable data throughput.

As you will have gathered from the discussion in section 2.2, a maximum data rate of 5 Mbyte/s may be achieved on the 8-bit wide SCSI bus if asynchronous data transfer is used, without having to recourse to reduced Fast-SCSI timing. This transfer rate alone could clearly compromise the ISA bus (2 to 3 MByte/s). If, however, Fast-SCSI compatible devices are used, and an appropriate adapter, the ISA bus presents an unacceptable bottleneck. So, if the PC offers a faster bus system than the venerable ISA, then a SCSI adapter should be bought for the faster bus (for example, PCI), if only to make sure that the performance of SCSI is not given away.

However, in general, it can not be said that running a SCSI adapter on an ISA bus makes no sense. If, for example, two adapters are used in one computer, say, because more than eight SCSI devices are to operate on this computer, and the user allocates the fast SCSI devices (hard disks etc.) to SCSI adapters having a PCI connection, then the data throughput offered by the ISA bus may be quite acceptable for the remaining, slower, devices (CD-ROM, tape drive, scanner, etc.). In this case, purchasing an inexpensive ISA-style SCSI adapter is definitely worthwhile. Alternatively, you may find good use for an older, already available, SCSI adapter.

With PCs running under Mac-OS (Apple and clones), this kind of symbiosis is not possible. For one thing, Apple nearly always uses on-board adapters on Nubus computers. Obviously, these adapters may not be integrated into a new system. On the other hand, a Nubus slot is no longer available in the new PCI computers.

Termination

Because it is so often done the wrong way during the configuration of the SCSI bus, the termination is one of the most important jobs. Termination involves fitting termination resistors at both cable ends of the SCSI bus.

Termination is required because electrical cables reflect high-frequency signals at the cable ends. The reflected signals are then superimposed on to the original ones (i.e., the one to be transmitted), which may be considerably distorted in this way. The reflection factor (for the current and voltage levels) is highest with open-circuited or short-circuited cable ends. If, however, the cable is terminated with its characteristic impedance, the reflections are reduced to a minimum. In SCSI terminology, such a cable termination resistance is called a *Terminator*. The passive version of the terminator is an array of 220/330-Ω resistors. Such a cable termination is shown in Figure 3.3. This should be present on each signal wire. Using this potential divider, the wire is held at a nominal voltage. The supply voltage marked *Termination Power* (TERMPWR = termination power) equals +5 V, so that the voltage

3.1 Configuration of the SCSI bus

divider keeps each line at a level of +3 V. This definition of the direct voltage level is also essential to prevent undefined levels straying across the bus during the *Bus-Free phase*, possibly causing an unintentional transition to another phase.

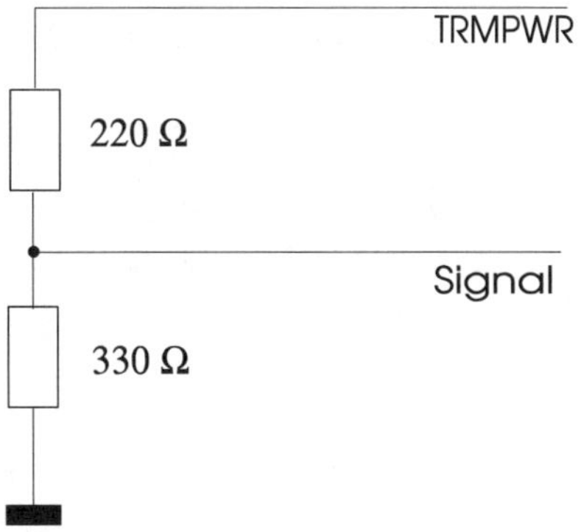

Even in asynchronous mode, but particularly with synchronous data transmission, the wires in a SCSI cable convey signals having frequencies in the megahertz (MHz) range. Consequently, the cable ends must always be terminated properly to guarantee error-free transmission.

Figure 3.3.
Passive termination of a signal wire on the SCSI bus.

A SCSI adapter offers an internal and an external connector. When internal as well as external devices are connected, termination must be effected on the last device (internal as well as external).

Where only internal or only external devices are available, termination must be effected on the last SCSI device and on the SCSI adapter, because the latter then forms the wire end of the bus.

The three possibilities are clarified by Figure 3.4:

a) both internal and external devices are connected;
b) only internal devices exist;
c) only external devices are available.

SCSI devices are always connected to form a chain, the two ends of which are formed by the terminators.

Although the rules for termination are logical and easy to comply with, errors are still made all the time. Not so much when a system is configured from scratch, but particularly when a device is added 'for the occasion', or removed from any location, and you forget to modify the termination accordingly.

3. SCSI in practice

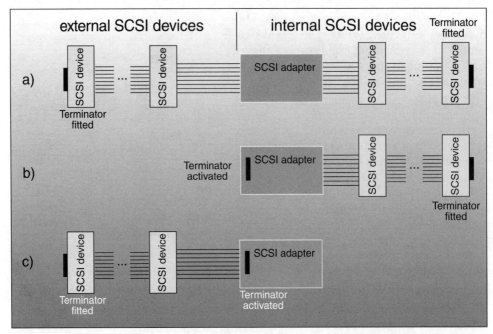

Figure 3.4.
Termination of the SCSI bus.

Hence, the rules for termination are given once again in emphasized form:

If internal as well as external SCSI devices are connected, the beginning and the end of the SCSI cable is at the last internal or external device respectively; these devices **must** therefore be fitted with a terminator.

If **only** external **or** internal devices are available, the beginning of the SCSI cable is at the adapter, and the end at the last device, in both cases. In this case (*b* or *c*) the termination resistors must be fitted on the SCSI adapter, and one terminator on the last device. If the reverse of situation *a* is applicable (internal *and* external devices available), then the termination resistors must be removed from the SCSI adapter board.

3.1 Configuration of the SCSI bus

Depending on the device type, termination may be effected in one of three ways:

- by fitting a resistor array (different casings);
- by fitting a jumper which connects or disconnects available resistors;
- by a software command, where a terminator (usually an active type) is switched on or off (often found on SCSI adapters).

Some versions of plug-in terminators are shown in Figures 3.5 and 3.6.

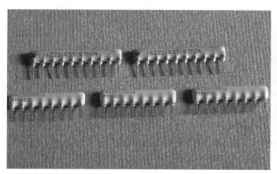

Figure 3.5.

Termination resistor arrays; their application is restricted to internal devices and SCSI adapters.

Figure 3.6.

Plug-on terminators for external devices (top) and internal cables (below, rarely used).

When termination errors occur, either too many terminators are used on the SCSI bus (for example, a SCSI device was added and properly terminated being the last device, but you forgot to switch off the termination on the device that used to be last in the chain), or a Terminator was forgotten at the end of the bus. In both cases, errors need not be noticed straight away. In the long term, however, they may have unpleasant consequences.

Case 1: One Terminator missing. From experience, this error is hardly noticed with devices having a relatively low data transfer rate (CD-ROM, slow removables, scanners). With fast hard disks,

135

however, read and write errors will occur frequently. A *SCSI Reset*, too, is often performed incorrectly. This is marked by a continuously lighting SCSI LED, which normally indicates access to SCSI devices by short flashes. In many cases, this condition may only be lifted by a hard reset of the computer.

Case 2: Two Terminators missing. This means that the bus is not terminated at all. In general, nothing will work if that is the case.

Case 3: One Terminator too many. This error occurs frequently when only internal SCSI devices used to be employed, and the computer is extended with an external device at a later stage. Although the external device is usually correctly terminated in these cases, you forget to remove the previously needed termination from the SCSI adapter.

The data transfer is rarely affected by the third terminator. Do remember, however, that this additional Terminator forms an extra load on the *TERMPWR* line because the Terminators are connected in parallel as far as the d.c. supply is concerned.

Case 4: Two Terminators too many. The error cause is the same as above: while extending the SCSI system you have forgotten to remove one or more redundant Terminators.

The cable is then terminated with only half the nominal resistance, which may cause difficult to reproduce errors particularly at high data speeds. It is possible for data loss to occur as a result of chained hard disk sectors. The load on the *TERMPWR* supply has been increased once again, and may cause overheating or even a burnout of the voltage regulator. A repair job is then unavoidable. The voltage on the *TERMPWR* line is usually supplied by the SCSI adapter board (on which the fuse or voltage regulator is likely to give up the ghost). Although the SCSI specification allows other SCSI devices to take over the *TERMPWR* supply, only **one** supply may be active at any time.

3.1 Configuration of the SCSI bus

Active terminators

Active termination based on a constant-current source and series resistors is an alternative to the previously discussed passive termination using a resistor array.

The advantage of active termination is that the voltage levels on the bus become load-independent, while parasitic capacitances are reduced. This is of particular significance for Fast-SCSI timing. The basic schematic of an active Terminator is shown in Figure 3.7, while an active Terminator for external devices is shown in Figure 3.8.

According to the SCSI specification, active and passive Terminators may be mixed, that is, a passive Terminator may be fitted at one cable end, while the other end receives an active Terminator.

As long as there are no problems, you need not worry about the *type* of termination: if the SCSI adapter has an active termination, you may still fit a passive termination at the other end of the bus.

Active termination is recommended whenever several devices are connected to the bus, and external cables are relatively long, approaching the maximum allowable cable length. Using an active Terminator then considerably improves pulse steepness and signal levels. Experience shows that trouble-ridden systems operating at high transfer rates, using long cables and passive Terminators,

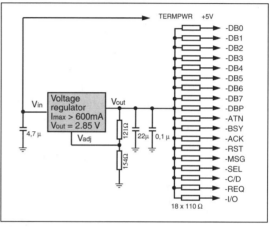

Figure 3.7.
Basic circuit diagram of an active terminator having a current source.

Figure 3.8.
An active terminator for external SCSI devices. The LED is the only distinctive feature.

3. SCSI in practice

are capable of running without problems if active Terminators are used. A condition is then, of course, that an active Terminator is used at both cable ends.

Note: the circuit diagrams of active as well as passive termination refer to the generally used single-ended SCSI system. Differential SCSI is discussed in an Appendix elsewhere in this book.

ID numbers

As already discussed in the theoretical section of this book, the maximum number of SCSI devices (including SCSI adapter) allowed on the bus is determined by the number of datalines. Figure 3.9 once again clarifies the relation between dataline and SCSI ID, which becomes significant during the Arbitration phase.

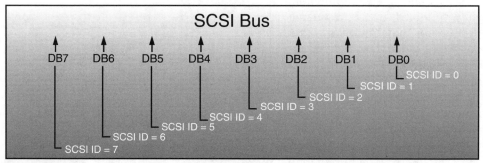

Figure 3.9.
Correlation between SCSI ID and available data lines.

As a result, a regular 8-bit wide bus allows seven SCSI devices and an adapter to be operated. More are allowed on a Wide-SCSI bus, depending on the bus width.

Any SCSI ID may be assigned once only to make sure that each device on the bus can be addressed using its own, unique identification. The ID number has to be set or changed by the user. The ease of setting an ID number varies between external and internal devices.

On internal devices, a jumper or DIP-switch setting is the most commonly used. For eight possible ID numbers, a block consisting of three jumpers or switches must be present ($2^3 = 8$). The manu-

3.1 Configuration of the SCSI bus

SCSI ID	Jumper connection / Switch position		
	Block 1	Block 2	Block 3
0	off	off	off
1	on	off	off
2	off	on	off
3	on	on	off
4	off	off	on
5	on	off	on
6	off	on	on
7	on	on	on

Table 3.1.
Relation between jumper (or switch) settings and SCSI ID.

facturer then determines the relations between jumper and switch positions and ID numbers (see Table 3.1).

External devices usually incorporate rotary switches on which a pointer has to be turned to the desired number.

Priority assignment

As you will recall from the theoretical sections in this book, the allocation of a SCSI ID also arranges the matter of priority on the bus. Priority means which SCSI device gets access to the bus during the *Arbitration phase*, when several devices want to become active at the same time. The highest priority is given to SCSI ID number 7, which is usually claimed by the SCSI adapter. That leaves IDs 0 through 6 for the remaining SCSI devices. In this system, ID 0 has additional significance. If the computer is to start up from a SCSI hard disk, certain conditions have to be satisfied (see *SCSI BIOS*). One of these is that the boot disk has to be preset for the so-called *Boot Target ID*. The preset for the *Boot Target ID* is usually number 0. On some systems, this preset may be altered, however. A hard disk from which the system is supposed to boot must be assigned this particular ID number.

3. SCSI in practice

Parity checking

To complete the series of settings you have to look at on a SCSI device, I wish to cover the subject of parity checking in the following paragraphs.

Parity checking is the only way of discovering errors that may arise during a transmission between Initiator and Target, by way of the SCSI bus. However, doubts may arise as regards the security of the applied method.

Functional principle: one line for parity checking is added to every eight data lines. So, four parity lines are used with 32-bit Wide-SCSI.

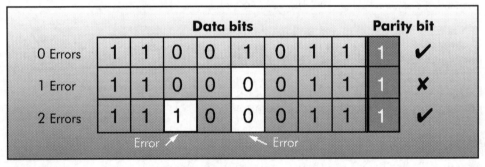

Figure 3.10.
Parity checking and its weak aspects.

The operation as well as the weakness of the parity checking method is illustrated in Figure 3.10. The checksum of the individual data bits is completed by the parity bit in such a way that the checksum of all nine bits is an even number (*even parity*).

If a data bit is corrupted in a transmission (in Figure 3.10, a 1 is turned into a 0), the checksum is no longer even-numbered, allowing the receiver to recognize that an error has occurred, and request a repeat of the transmission.

So far, so good. If, however, two errors occur simultaneously in a line, the checksum is even-numbered again, so that the errors are not detected. In other words, the parity checking method as described above is only capable of detecting an odd number of errors.

3.1 Configuration of the SCSI bus

Moreover, the parity bit itself may be corrupted: an otherwise correct transmission may then be rejected as faulty.

So, the results of a parity check are questionable, as is the purpose of this method.

Parity checking used to be applied in the working memory (RAM) fitted in PCs. After the introduction of PS/2-style SIMMs, which were supplied without a parity bit, parity checking became obsolete. It should be noted, though, that the conditions inside a PC (having a closed metal case) may not be likened to those that apply to a data transmission via external cables. In cables, the risk of interference is certainly higher, increasing the probability of a data transmission error. However, the assumption that parity checking is the right instrument for error detection remains questionable.

Figure 3.11.
Switches on the rear panel of an external device – the rotary switch at the top is used to set the SCSI ID, while the first DIP switch in the lower block sets the parity checking.

I started this short discourse in the first place to discuss the required configuration settings. Many SCSI devices have a jumper or a DIP switch which allows the user to select between operation with or without parity checking. Note, however, that the yes/no selection you make in this respect has consequences for the entire bus. The rule of the game is: all or none at all. If parity checking is deemed necessary, it has to be actuated on all devices on the bus. Likewise, if it is not needed, it has to be switched off on all devices.

As a result of this rule, you have to decide at an early stage on applying or omitting parity checking on your SCSI devices. Also, when the system is extended, you have to make sure that the parity on/off setting on the new device is in accordance with the choice you made earlier. Complications may arise if you do not observe this simple rule, and mix devices using parity checking with non-checking ones in a single system.

Even if parity checking is by no means a certain method of detecting transmission errors, it is fairly reliable when it comes to detecting gross errors caused by major interference sources (for example, an entire data line is corrupted to such an extent that it is noted by the parity check).

As long as you are not forced to switch off parity checking on all devices including the adapter (for instance, if a SCSI-1 device without parity checking has to be connected to the bus), it is recommended to operate the SCSI bus using parity checking (*Parity enabled*).

SCSI adapter installation

Essentially, there is no difference between the installation of a SCSI adapter and any other insertion card. I do not wish to go into details here, because there is no such thing as a general-purpose mounting instruction for SCSI adapters. The installation procedure depends on the computer type used (*IBM compatible, Apple, Risc computer, Amiga*, etc.), as well as on the operating system (*DOS, Windows 3.1, Windows 95, MacOS, Unix, ...*), on the bus system available in the computer which is to work with SCSI (*PCI, ISA, MCA, EISA, VL, NuBus, ...*) and, last but not least, on the plug & play capability of the adapter. More detailed information on possible installation problems with each card will be given further on in this chapter with my presentation of SCSI adapters.

In the case of Windows/Intel computers, which have by far the largest share of the PC market, the following rule applies: one free interrupt, one free DMA channel and one free base address are required. If these are available, nothing can come in the way of successful installation, provided, of course, you keep to the installation directions supplied by the relevant manufacturer.

In case the termination of the adapter is software-controlled, the SCSI adapter may be plugged into the desired bus slot (the computer being disconnected from the mains, and the user having taken sufficient precautions against static discharges).

If, on the other hand, the SCSI adapter has to be terminated manually, that is, by setting or removing a jumper, or removing resistor arrays, then some thought should be given to the configuration of the future SCSI system.

If only internal SCSI devices are available, the adapter must be terminated. If there are internal as well as external devices, the termination should be removed in any case. These actions should be finished before the card is inserted into the slot. Once the card has been inserted, removing resistor arrays or swapping a jumper seems to require an extra wrist in the lower arm.

Particularly annoying in such cases is the decision in favour of "*ah yes, maybe ...*". The indeterminate among you may be advised to buy an external Terminator that fits on the external SCSI-2 connector of the adapter (Figure 3.12). The termination on the card is then removed in any case, and the active Terminator is plugged on to the external SCSI connector if no external devices are used. If you do decide to use external devices, the Terminator is removed, and the device(s) is (are) connected via a cable. The Terminator is then moved to the last device in the chain, which is so provided with the required bus termination. No modifications should then be undertaken anymore on the SCSI adapter proper.

Figure 3.12.
An external terminator may help to decide whether or not the SCSI adapter has to be terminated.

At the internal side, the decision is rather easier: a SCSI hard disk or a CD-ROM drive is usually available so that here, too, the last device in the chain is terminated in accordance with the rules.

3. 2 SCSI cables

Cables are often a neglected aspect of the SCSI system. With increasing data speeds, there is an increased likelihood of bits being corrupted on the bus as a result of external interference when cables with inadequate screening are used. Everything internal to the computer is relatively uncritical in this respect. Firstly, the metal case of the computer is a good protection against external inter-

3. SCSI in practice

ference. Secondly, flatcables with alternate signal/ground wires are applied to connect internal SCSI devices.

By contrast, round cables are used to connect external devices. The effect of noise sources on these cables should be ascertained with a far more critical eye. For one thing, noise signals may enter the cable directly, on the other hand, it is difficult to see from the outside if the cable has sufficient screening, low-cost manufacturers having discovered the market for external SCSI cables long ago.

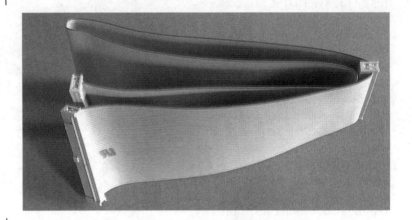

Figure 3.13. Flatcable with plugs for internal SCSI connections, 50-way.

Internal SCSI cables

Figure 3.13 shows a typical flatcable as used to interconnect internal SCSI devices. Normally, three insulation-displacement connectors (IDCs) have been pressed on to the cable, allowing it to interconnect several internal devices.

Table 3.2 once again clarifies how the alternate distribution of the ground wires across the 50-way cable can help to provide sufficient screening.

Pin 1 is usually marked (triangle, arrow), while the associated wire often has a different colour (red, black or blue side) from the rest of the cable. A polarizing bar in the 50-way socket and a matching notch in the connector are provided to ensure that the two can not be connected the wrong way around. There are, however, a few odd SCSI devices and adapters on which it is possible to fit the socket the wrong way around. If that is the case, take good care

3.2 SCSI cables

to align the pin marked '1' on the socket with the corresponding pin on the plug, else, short-circuits are likely to occur on the SCSI bus, or other conditions causing damage.

With increasing data rates, flatcables can not guarantee entirely noise-free data exchange, particularly when *Fast-SCSI* is used. Because of this risk, a minimum distance of 1.3 mm should be observed inside the computer between the SCSI flatcable and other cables, metal parts and, in particular, folded sections of the same cable (cross-talk). This minimum requirement should not be too difficult to satisfy.

In *Wide-SCSI* mode, an additional 68-way cable is used for the connection between adapter and devices. Such a cable is connected via a high-density 68-pin plug. The relative size of this plug as compared with the regular 50-way type is illustrated in Figure 3.14.

Signal name	Plug pin		Signal name
GROUND	1	2	-DB(0)
GROUND	3	4	-DB(1)
GROUND	5	6	-DB(2)
GROUND	7	8	-DB(3)
GROUND	9	10	-DB(4)
GROUND	11	12	-DB(5)
GROUND	13	14	-DB(6)
GROUND	15	16	-DB(7)
GROUND	17	18	-DB(P)
GROUND	19	20	GROUND
GROUND	21	22	GROUND
RESERVED	23	24	RESERVED
FREE	25	26	TERMPWR
RESERVED	27	28	RESERVED
GROUND	29	30	GROUND
GROUND	31	32	-ATN
GROUND	33	34	GROUND
GROUND	35	36	-BSY
GROUND	37	38	-ACK
GROUND	39	40	-RST
GROUND	41	42	-MSG
GROUND	43	44	-SEL
GROUND	45	46	-C/D
GROUND	47	48	-REQ
GROUND	49	50	-I/O

Table 3.2.
Contact assignment on the 50-pin internal SCSI plug (single ended).

The 50-way cable is the de-facto standard under SCSI-2, and is referred to as the *A-Cable*. The 68-way cable that allows the transmission width to be widened to 16 or 32 bits is called *B-Cable* (meanwhile replaced by the P-Cable with 16-bit Wide-SCSI, see page 153).

The following typical characteristics have been defined for both cables when Fast-SCSI timing is used:

- impedance: 90 to 132 Ω
- max. wire resistance at 20°C: 0.23Ω/m
- attenuation: 0.095 dB/m at 5 MHz
- propagation difference between two wire pairs: < 0.2 ns/m

These typical values also apply to external (round) cables.

3. SCSI in practice

Figure 3.14.
The 50-pin internal SCSI plug (right) and its 68-pin partner for use in Wide-SCSI systems.

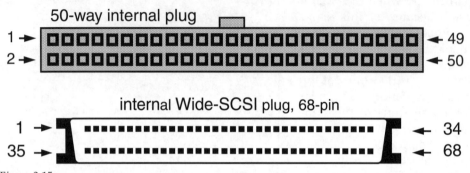

Figure 3.15.
Pin functions of both plug types.

External SCSI cables

While the SCSI cabling inside a computer is realized with flatcables and uniform plug/socket types, things look rather different at the external side.

External cables suffer from far greater exposure to interference than internal ones. Because of this, the use of *twisted-pair cables* with good screening is recommended (each signal cable is twisted with an associated ground wire). Unfortunately, these are not exactly cheap. Consequently, users will try to save money, after all, who is prepared to fork out £20 or so for a short cable to connect two peripheral devices? Here, again, the low-cost industry has a sales opportunity, offering cables at roughly half the price. In practice, however, most of these cables are not cost-effective solutions but cheap only and of dubious quality. The twisted-pair principle is never applied in these cheap cables, and a single screening simply serves to keep the cable together.

3.2 SCSI cables

Signal name	Pin number		Signal name
GROUND	1	35	GROUND
GROUND	2	36	-DB 8
GROUND	3	37	-DB 9
GROUND	4	38	-DB 10
GROUND	5	39	-DB 11
GROUND	6	40	-DB 12
GROUND	7	41	-DB 13
GROUND	8	42	-DB 14
GROUND	9	43	-DB 15
GROUND	10	44	-DB P1
GROUND	11	45	-ACKB
GROUND	12	46	GROUND
GROUND	13	47	-REQB
GROUND	14	48	-DB 16
GROUND	15	49	-DB 17
GROUND	16	50	-DB 18
TERMPWR	17	51	TERMPWR
TERMPWR	18	52	TERMPWR
GROUND	19	53	-DB 19
GROUND	20	54	-DB 20
GROUND	21	55	-DB 21
GROUND	22	56	-DB 22
GROUND	23	57	-DB 23
GROUND	24	58	-DB P2
GROUND	25	59	-DB 24
GROUND	26	60	-DB 25
GROUND	27	61	-DB 26
GROUND	28	62	-DB 27
GROUND	29	63	-DB 28
GROUND	30	64	-DB 29
GROUND	31	65	-DB 30
GROUND	32	66	-DB 31
GROUND	33	67	-DB P3
GROUND	34	68	GROUND

Table 3.3.
Pin assignment of the 68-way B-cable (single-ended SCSI). From the fact that only datalines DB8 through DB31 are available you may deduce that this cable is actually an extended version of the A-cable, although this is not replaced by it.

Unfortunately, the poor quality of these round cables is not immediately apparent from the outside. At best, a certain flexibility of the cable may be an indication that you are holding a low-cost product.

Like cables, connectors are also subject to quality differences. In most cases, HD (*high density*) plugs are used to connect cables to SCSI equipment. For regular 8-bit SCSI, 50-way HD plugs and

3. SCSI in practice

Figure 3.16.

Cross-sectional view of a low-cost round-cable. The individual wires are not twisted, and a screening braid is missing. The only screening is a single aluminium layer, featured in better cables as an addition to a real screening braid.

mating sockets are used. For Wide-SCSI, on the other hand, a 68-way version is employed. Both versions are shown in Figure 3.17.

At the equipment side, the new HD connections are not yet in wide use; in most cases, you will still find 50-way Centronics-style connections as shown in Figure 3.18, at least for 8-bit SCSI devices (external Wide-SCSI devices being rare birds as yet).

While it is true that the Centronics plug/socket combination is the more robust, a 68-pin Wide-SCSI connection would have been too bulky in this style. Because of this, the far more compact HD connections were developed in 50 and 68-way versions, of course.

The link between the SCSI adapter and the first external device requires an *HD-to-Centronics* cable, while the link between two external devices is made with the aid of a cable having Centronics plugs at both ends. Technically, the two connection types are equal, and any computer shop should be able to supply HD-to-Centronics and Centronics-Centronics cables from stock.

With Centronics as well as HD cables, the connector on the equipment is a socket, and the part on the cable is a plug. The advantage of this arrangement is that no repairs are required at

Figure 3.17.
Hd connections for Wide- and 8-bit SCSI.

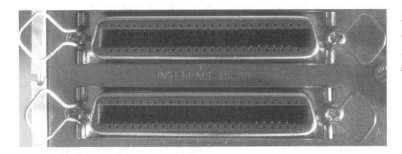

Figure 3.18. Centronics connectors on external devices.

the side of the SCSI equipment if the plug is damaged (bent pins, etc.) — just a new cable. This precaution makes sense because connector wear and tear must be taken into account with external equipment which should be easy to migrate between computer systems.

A third connector type, the 25-pin *sub-D* style, is (unfortunately) still in use for external adapter connections. Apple in particular equip their computers with such external adapter connectors.

The use of only 25 pins may cause raised eyebrows, and can only be explained by the simple omission of a large number of ground connections. This is confirmed by Apple's technical documentation and the pinning of the relevant sub-D plug (Figure 3.19). Only six ground wires are left of the 26 originally designed into the 50-way cable.

Obviously, the absence of the required number of ground wires increases the noise susceptibility of an external SCSI cable.

Apple seems to be aware of problems in this respect, the *Tech Info Library* providing the following statement to this effect:

In case the cable length across all devices exceeds 10 feet (3 m), the cable should be terminated at the 10-feet point, in addition to the termination at each cable end. If more than three devices are operated on the bus, cable lengths of 3 m are often exceeded, requiring three terminators for the system to operate correctly. Do not add a third terminator as long as problems are not apparent, or have a different cause.

3. SCSI in practice

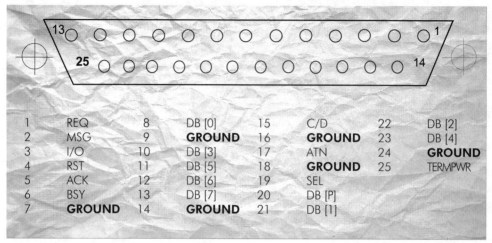

1	REQ	8	DB [0]	15	C/D	22	DB [2]
2	MSG	9	**GROUND**	16	**GROUND**	23	DB [4]
3	I/O	10	DB [3]	17	ATN	24	**GROUND**
4	RST	11	DB [5]	18	**GROUND**	25	TERMPWR
5	ACK	12	DB [6]	19	SEL		
6	BSY	13	DB [7]	20	DB [P]		
7	**GROUND**	14	**GROUND**	21	DB [1]		

Figure 3.19.
Contact assignment on the 25-pin sub-D plug (Apple) having just six ground pins.

Because the 'Apple-standard' sub-D/Centronics cables I had available were relatively stiff as compared with low-cost versions, the question arose how a 50-way cable may be connected to a 25-way plug in a sensible way. Cutting the cable revealed a twisted-pair cable with proper screening, although only 38 of the expected 50 wires were available. Because a SCSI cable should contain 18 signal wires plus one for TERMPWR, the number of wires appears to be just sufficient to twist each signal wire with its own ground wire. But how do you connect 19 ground wires to just 6 plug pins?

Measurements with my ohmmeter revealed that several ground wires were joined on one ground pin. Although this is definitely better than leaving them open-circuited, this commoning of ground wires does introduce the risk of earth loops. Glitches will be amplified to some extent on the cable.

Apple should decide to eliminate the occasionally occurring problems with external SCSI devices by applying external HD-style plugs for the on-board SCSI adapter, instead of attempting to combat the symptoms with a third terminator on the SCSI bus. Apple users wishing to avoid such problems in the first place have no alternative but to use an additional insertion-card style SCSI adapter which should offer an external HD connector.

3.2 SCSI cables

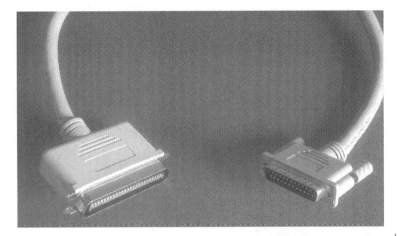

*Figure 3.20.
A sub-D/Centronics cable to Apple standards; the problem is the missing ground pins at the side of the sub-D plug (right).*

Not relating to these problems with a special cable type I would like to make two further remarks on the subject of external SCSI cables:

- Do not go for low-cost alternatives if you buy external SCSI cables. Their poor screening causes considerable problems particularly at higher data transfer rates.
- External SCSI cables should always be as short as possible to reduce the impact area for external noise sources. This also applies if the maximum allowable total cable length of 6 m (or 3 m with Wide-SCSI) has not been reached. The standard length of a cable between two external SCSI devices is 30 cm.

Cable quality under Fast-SCSI

The quality of external round cables becomes more important particularly at high data rates and reduced timing schemes. Capacitive effects and cross-talk may badly affect reduced timing. Signal edges, for example, may be become so slow that proper interpretation becomes impossible.

Furthermore, only good-quality connectors should be used, as well as active Terminators (because of their smaller parasitic capacitance), and cables having an equal impedance. Even the arrangement of the wire pairs in the cable becomes significant, because in round cables the impedance of the inner wire pairs is higher than that of the pairs at the outside of the cable.

3. SCSI in practice

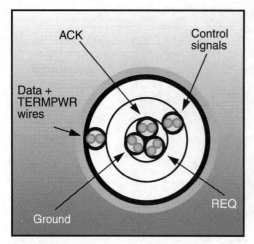

Figure 3.21.
Wire pair arrangement in a round cable, which is a must for Fast-SCSI.

Therefore, the *ACK* and *REQ* lines should run in the centre of the cable, while *TERMPWR* and the datalines should go to the outer layers (see Figure 3.21). In this arrangement, an individual ground wire is essential for each signal wire.

The 25-way sub-D plug as applied by Apple should not be used under any circumstances.

Cable standards for SCSI-3

A and B-cables will no longer appear in the new, as yet unfinished, SCSI-3 standards. The intention is to use 68-way cables only. The standard cable is to be called *P-Cable* and may be used for transmission widths of up to 16 bits. The *Q-Cable*, which is also a 68-way type, then forms a complement to the P-Cable, the two allowing a data transmission width of 32 bits. The P-Cable conveys datalines DB0 through DB15, while the Q-Cable is responsible for the upper half of the datalines (DB16 through DB31).

If you compare Table 3.4 with Table 3.3, you will notice that the 68-way B-Cables used so far can guarantee Wide-SCSI transmissions in conjunction with A-Cables only, while a P-Cable on its own allows a 16-bit Wide-SCSI connection to be implemented.

In view of the ever higher data rates, cable quality will become an increasingly prominent issue, particularly with SCSI-3.

Plug connections

The minimum and maximum values for cable parameters include the entry *Plug connection: maximum 10 cm*. This value applies as a maximum for the addition of all plug connection lengths on the SCSI bus. It will be appreciated that the value is relatively small. But what is a plug connection?

3.2 SCSI cables

Signal name P + Q cables	Pin number		Signal name P cable	Signal name Q cable
GROUND	1	35	-DB 12	-DB 28
GROUND	2	36	-DB 13	-DB 29
GROUND	3	37	-DB 14	-DB 30
GROUND	4	38	-DB 15	-DB 31
GROUND	5	39	-DB P1	-DB P3
GROUND	6	40	-DB 0	-DB 16
GROUND	7	41	-DB 1	-DB 17
GROUND	8	42	-DB 2	-DB 18
GROUND	9	43	-DB 3	-DB 19
GROUND	10	44	-DB 4	-DB 20
GROUND	11	45	-DB 5	-DB 21
GROUND	12	46	-DB 6	-DB 22
GROUND	13	47	-DB 7	-DB 23
GROUND	14	48	-DB P	-DB P2
GROUND	15	49	GROUND	GROUND
GROUND	16	50	GROUND	GROUND
TERMPWR	17	51	TERMPWR	TERMPWR
TERMPWR	18	52	TERMPWR	TERMPWR
reserved	19	53	reserved	reserved
GROUND	20	54	GROUND	GROUND
GROUND	21	55	-ATN	terminated
GROUND	22	56	GROUND	GROUND
GROUND	23	57	-BSY	terminated
GROUND	24	58	-ACK	-ACKQ
GROUND	25	59	-RST	terminated
GROUND	26	60	-MSG	terminated
GROUND	27	61	-SEL	terminated
GROUND	28	62	-C/D	terminated
GROUND	29	63	-REQ	-REQQ
GROUND	30	64	-I/O	terminated
GROUND	31	65	-DB 8	-DB 24
GROUND	32	66	-DB 9	-DB 25
GROUND	33	67	-DB 10	-DB 26
GROUND	34	68	-DB 11	-DB 27

Table 3.4.
Pinouts of the new P- and Q-cables designed for SCSI-3.

A plug connection is the (shortest possible) connecting wire between the tap on the SCSI bus and the actual device plug. The maximum value of 10 cm is the reason for the use of two connectors on all external SCSI equipment: one for the input plug, and another to take the bus further, or for termination.

The cable arrangement is illustrated in Figure 3.22: inside the enclosure, the two SCSI sockets on the rear panel are connected by a length of flatcable to which the actual device plug is attached. The length of the plug connection — the distance between the tap

Figure 3.22.
Cabling for SCSI devices in an external case.

on the cable and the contacts on the SCSI device — is, in general, a few millimetres only to make sure that the total length of all plug connections does not exceed the maximum value of 10 cm even if the highest possible number of devices is connected to the SCSI bus.

3. 3 Software interfacing

SCSI is a universally applicable system. Devices of widely different classes may be operated on the SCSI bus, and host adapters are available for virtually any computer type and any computer bus. The data exchange between SCSI devices, as well as between device and SCSI adapter, operates without problems. But how does the communication between operating system and SCSI host adapter work?

The software bridge between the SCSI adapter and the operating system should comply with the following requirements:

- Any SCSI adapter of any manufacturer, for which a software driver is available for the desired operating system, may be used.
- Any SCSI device from any manufacturer may be addressed via the driver supplied by the adapter manufacturer, or via a driver supplied by the device manufacturer.

These two conditions describe the circumstance that either the adapter manufacturer is to supply drivers that work with a SCSI device which may be totally unknown to the device programmer, or, the other way around, the device manufacturer supplies a

3.3 Software interfacing

driver for his device, this driver being able to establish the connection to the operating system via any SCSI adapter.

All of this calls for an interface with clear definitions, and offering provisions for both sides that allow such a software driver to be programmed without any knowledge being available on the device or adapter at the other side of the interface.

Two such interfaces are available: CAM (*Common Access Method*) and ASPI (*Advanced SCSI Programming Interface*). These lighten the work of programmers writing drivers for these software interfaces thanks to the fact that all devices belong in certain SCSI classes which may be addressed with a uniform set of commands.

CAM (Common Access Method)

The CAM interface was defined by the ANSI Committee and offers a standardized procedure which helps to address widely different SCSI devices via widely different SCSI adapters. It is based on two types of driver — the CAM driver which creates the manufacturer-specific connection to the SCSI adapter, and drivers for the relevant device class, which are responsible for the link to the operating system and the applications.

Essentially, the CAM driver consists of two parts:

- **Transport Module XPT**
  This module diverts device access operations from applications to the SCSI adapter(s), and also arranges for desired data to travel from peripheral devices to the program which requests them.

- **Interface Module SIM** (*SCSI Interface Module*)
  The Interface Module is in control of the resources of the SCSI adapter, and forms a device-independent interface for drivers and applications. In other words, it builds the link to the XPT module.

The way in which the CAM driver is embedded and the actual sequence of the initialisation are strongly dependent on the operating system. Under DOS, the practical situation is that a so-called DOSCAM driver is loaded (access to SCSI adapter), to which drivers are attached for the relevant SCSI class.

If an application wants to address a SCSI device, this device access operation is conveyed via the following stations:

Application/Operating system—Device class driver—CAM driver—SCSI adapter—SCSI bus—SCSI device.

In this way, the CAM driver and the associated device driver form a bridge between operating system and SCSI adapter. Any user purchasing, for example, a SCSI adapter for a Windows/Intel computer, but not quite sure of the operating system which is to run on the PC shortly (DOS, Windows NT, OS/2, etc.), is well advised to ascertain that the SCSI adapter comes with CAM drivers for all operating systems. This is the only way to be armed against any eventuality in this respect.

Device drivers

A device driver has the following tasks:

a) interpret I/O requests from applications or the operating system;
b) divert such requests to the XPT/SIM Control Blocks;
c) provide required resources, *CAM Control Blocks* (CCB) as well as buffers;
d) manage exception conditions (e.g., an unexpected Bus-Free phase, etc.);
e) record and file such exception conditions for the purpose of supporting analysis tools;
f) carry out formatting jobs, or lend assistance with these;
g) hold ready transfer parameters for the SCSI adapter;
h) guide SCSI access operations to the right path/bus, Target or LUN;
i) take over configuration functions for a Target which are not run during initialisation or formatting;
j) provide time-out values for CCBs.

3.3 Software interfacing

XPT functions

The XPT module has the following tasks to contend with:

a) connect the CAM Control Block to the right SIM module;
b) manage CCB resources;
c) manage device class parameters;
d) with asynchronous operations, return handshake acknowledgement to the SCSI device.

SIM functions

The SIM module has the following tasks:

a) take over the function of interfacing to the SCSI adapter;
b) execute the individual SCSI protocol steps, or delegate them;
c) initiate error correction measures when errors occur;
d) control the data transfer to the adapter;

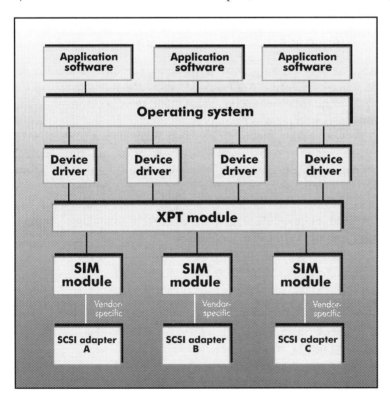

Figure 3.23. Basic structure of the Common Access Method.

e) arrange operations for one LUN or several LUNs in the right order, if necessary insert them into queues;
f) freeze and enable queues;
g) send confirmation of completed operations to the device driver;
h) handle device selection, device connection and device interruption, as well as data pointer management;
i) control a time function using requirements supplied by the device driver.

Individual levels of the *Common Access Method* are shown in Figure 3.23. If you require more detailed information on the CAM interface, possibly with the aim of programming your own device driver, I recommend referring to the file *CAM-R12B.PDF* which may be found on the CD-ROM supplied with this book. In addition to small programming routines intended to clarify the individual sequences, this file also provides detailed information on the embedding of the CAM interface into different operating systems.

ASPI (Advanced SCSI Programming Interface)

ASPI follows the same concept as CAM. The order should really be reversed, because ASPI existed before the introduction of the CAM standard. The leading company in the development of this software interface was Adaptec.

As with CAM, the bridge between SCSI adapter and SCSI device is formed by two drivers which build on one another: an ASPI Manager, which comes with the relevant SCSI adapter and observes its hardware peculiarities, and driver modules which are geared to the relevant device class and its command set. Just as a CAM driver, an ASPI manager is capable of managing I/O operations on the various SCSI device classes.

Seven ASPI functions have been defined:

- Host Adapter Inquiry
- Get Device Type
- Execute SCSI I/O Command
- Abort SCSI Command
- Reset SCSI Command
- Set Host Adapter Parameters
- Get Disk Drive Information

3.3 Software interfacing

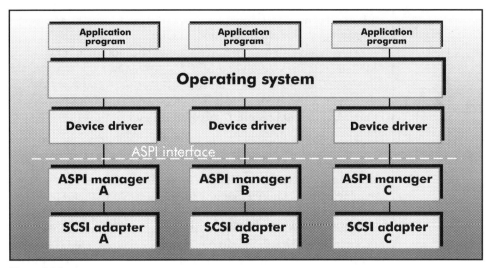

Figure 3.24.
Individual elements and levels of the ASPI interface.

The access of a device driver on a SCSI manager is performed in two steps. First (and once only), a so-called *Entry Point* is requested, whereupon the ASPI manager is called via this *Entry Point* (if necessary, several times).

If you want to know exactly how the *Entry Points* work, and how the calls are processed under the operating system of your choice, or if you want to write your own ASPI device driver, then be advised to have look at the files *ASPI_DOS.TXT*, *ASPI_OS2.TXT*, and others on the CD-ROM supplied with this book. Using examples, these files explain the way in which ASPI may be integrated into the different operating systems.

The graphic representation of the individual elements and levels of the ASPI concept shown in Figure 3.24 indicate the similarity between CAM and ASPI as far as their basic operation is concerned.

Because of the market dominance of the Adaptec company and the fact that ASPI was available before CAM, the former software interface is the more frequently found till now. A company like Symbios Logic (formerly NCR), however, emphasises the CAM interface and supplies a so-called ASPICAM driver with its SCSI adapters. This driver enables ASPI device drivers to access the CAM interface of these adapters as well.

In such a construction, however, one additional driver must be loaded (ASPI device diver, ASPICAM driver and the actual CAM driver). The cause of the addition is not only the ASPI device driver, since the driving of SCSI device classes functions just as well via CAM drivers. Rather, additional, auxiliary, programs are meant.

There exist a respectable number of SCSI utilities; small auxiliary programs which either originate from the shareware circuit, or are available free of charge as freeware. Occasionally, device or adapter manufacturers supply such utilities with their hardware. The utilities allow you, for example, to check the SCSI setup, or enable write caches on hard disk drivers. Alternatively, they may be monitor programs that enable individual SCSI commands to be put on the bus to check the operation of the addressed device. Most of the utilities have one thing in common, however: they are based on the ASPI interface.

The graphic representations of the CAM and ASPI concepts shown in Figures 3.23 and 3.24 indicate that the use of multiple, different, SCSI adapters under the two interfaces calls for CAM drivers or ASPI managers to be employed that belong with the relevant adapter.

Details on the installation of CAM and ASPI drivers may be found in the discussion of SCSI adapters from Symbios and Adaptec, whose Symbios 8250S, AHA-2920 and AHA-2940UW are presented in section 3.5 as representative for many other types.

3. 4 SCSI BIOS

At least SCSI hard disks can make do without drivers, and, consequently, without a *CAM* or an *ASPI* interface. Provided certain conditions are satisfied, these disks may be used with the support of the *SCSI* adapter BIOS (ROM).

This makes sense, because otherwise a computer equipped with a SCSI hard disk would only be able to start up via diskette or a non-SCSI disk (E-IDE, or similar). Remember, a computer is unable to start up from a hard disk to which it is 'attached' by drivers, because these drivers are loaded by the configuration files. Both the drivers and the configuration files are, however, on the hard disk, which should, of course, be made accessible beforehand.

A condition for the BIOS driving of a SCSI disk drive is a properly functioning co-operation between the computer and the SCSI BIOS of the adapter. If the PC is to start up from the SCSI disk drive, the computer may not contain an *(E-)IDE* disk which would be given priority.

The start-up (booting) sequence of a PC is a relatively simple sequence:

- First, RAM, system and similar checks are performed;
- next, the system looks for (E-)IDE disks; if found, these are copied into the CMOS list.
- Next, the system looks for extension cards of any type containing their own BIOS; these cards include SCSI adapters. A SCSI adapter then has the opportunity to copy the hard disk(s) it has to manage to the CMOS list (**after** any E-IDE disk that may be available).
- To complete the process, the boot sector of the disk which appears first in the list is given control of the system.

If no (E-)IDE disk is found during the start-up procedure, but only a SCSI drive, then the computer boots from the SCSI drive, because it is number 0 in the list.

Because SCSI is the standard interface on Apple computers (at least, till now), and the Apple BIOS recognizes SCSI disks immediately, these systems happily boot from SCSI disks.

To enable an (E-)IDE disk to be found during the start-up process, its identity has to be entered beforehand in the CMOS setup. If that is not done, the disk is not included in the boot list, and a SCSI disk would appear first. Depending on the operating system used, an (E-)IDE disk whose entry was removed from the CMOS list, but which is still present in the system, is or may be detected during the boot-up procedure, and may be used. Just as likely, however, the system locks up, or other problems arise.

3. SCSI in practice

No general indications may be given in this respect. If you want to start up from a SCSI disk, but at the same time wish to keep an (E-)IDE disk in the system, you can only keep trying, knowing that the chances of success are minimal. If the operating system is DOS, there is not even a small chance to succeed — DOS uses BIOS functions for the control of (E-)IDE disks, which makes it impossible to bypass such a disk during start-up.

Consequently, the generally valid steps to activate a bootable SCSI disk are as follows:

- Remove any available (E-)IDE disk(s) from the system, and clear the CMOS entry for this drive.
- Actuate the adapter's SCSI BIOS. On SCSI extension cards, that is usually accomplished with the aid of a jumper. If the system has an on-board SCSI adapter, the SCSI firmware may generally be actuated via the PC's CMOS setup.

Having completed these steps, the disk(s) should be recognized by the BIOS during the boot-up sequence, and the corresponding texts should appear on the screen, provided, of course, the SCSI adapter is properly installed, and the SCSI bus terminated in accordance with the familiar rules.

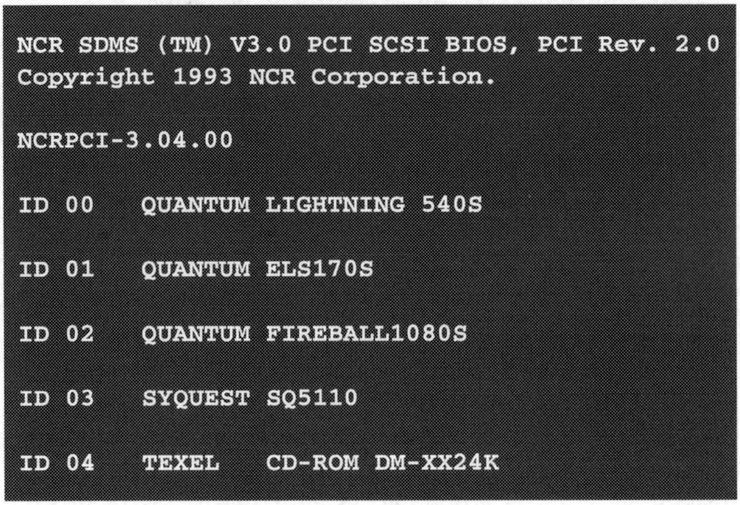

Figure 3.25.
Boot report produced by an ASUS board having a Symbios (previously NCR) on-board adapter. Three SCSI hard disks, a removable disk drive and a CD-ROM drive are recognized.

```
NCR SDMS (TM) V3.0 PCI SCSI BIOS, PCI Rev. 2.0
Copyright 1993 NCR Corporation.

NCRPCI-3.04.00

ID 00     QUANTUM LIGHTNING 540S

ID 01     QUANTUM ELS170S

ID 02     QUANTUM FIREBALL1080S

ID 03     SYQUEST SQ5110

ID 04     TEXEL    CD-ROM DM-XX24K
```

To make thing absolutely clear: (E-)IDE and SCSI disks may continue to sit in the PC, and work side by side in perfect harmony. In that case, the computer boots from the (E-)IDE disk, and not from the SCSI disk.

SCSI BIOS boot helper

Why is it that an IDE drive is bootable without any additional help, while a SCSI drive can not do without the extra BIOS provided by the adapter?

SCSI is a bus system for the connection of peripheral devices which is not bound to a certain computer type — SCSI should be able to work with the widest range of computers, from mainframes via workstations down to PCs. It is, therefore, necessary as well as useful to ensure that SCSI is not subject to any limitations whatsoever which so happen to be part and parcel of specific computer systems.

Int 13h

The best example is the *Int13h* hard disk interface of the PC/AT BIOS. When IBM originally defined this interface in 1986, the normal hard disk capacity was about 20 MBytes, or 60 MBytes for large types. It was assumed that hard disks in future would not require more than 1,024 cylinders. Consequently, only 10 bits were reserved for the CYL parameter of the Int13h interface (2^{10} = 1,024). Additionally, parameters for the number of heads and sectors were laid down.

Under SCSI, this fixed cylinder/head/sector allocation became a problem. SCSI addresses logic blocks and not cylinders, heads or sectors. Therefore, an algorithm had to be devised which allocates the CYL/HEAD/SECT parameters of Int13h to the logic SCSI blocks. Such a transformation is part of any disk attachment under SCSI, irrespective whether accomplished via the SCSI BIOS or a driver, and it explains why SCSI hard disks are not bootable without BIOS assistance.

As with many aspects of SCSI-1, the allocation was left to the manufacturers to resolve. Most of them decided to select the algorithm in such a way that each disk cylinder was allocated a memory capacity of one megabyte. The maximum number of 1,023 cylinders allowed by Int13h thus results in a maximum disk storage capacity of 1.023 GByte. At that time, the capacity was beyond imagination; today, it is nothing special. As a matter of course, the ever growing hard disk capacity forced manufacturers to let go of the fixed transformation size of 1 Mbyte per cylinder; a change of the mapping system became unavoidable.

Unavoidable, but painful. Although the correction of the transformation sizes does not restrict the practical use of SCSI hard disks with different relations between Int13h parameters and logic SCSI blocks, a disk using the old algorithms can not be read at first if the new mapping is applied — it has to be formatted!

In the mean time, algorithms are in use that allow a maximum disk capacity of 4 or even 8 Gbyte. Although this is certainly sufficient for the moment, the 8 Gbyte limit is sure to be reached in the foreseeable future if you extrapolate today's disk volume growth. Here, however, the combination PC BIOS/DOS operating system has a pitfall in store — more than $16 \cdot 10^{16}$ sectors are not allowed. If 512 bytes per sector are applied, the storage volume of SCSI hard disks can not grow beyond 8 Gigabyte under DOS (*IDE* disks reach the limit at 1 Gigabyte already, while the maximum capacity of *E-IDE* drives is currently 2 Gigabyte). Hopefully, modern operating systems will be raised to standards, and we will be able to dispose of ancient DOS before the 8-GByte limit starts to become a serious source of problems.

However, even today the use of different adapting algorithms has a few annoying aspects in stock. Problems lurk if a SCSI disk drive is operated in an external case with the obvious aim of being able to migrate the device quickly between different computer systems on which it may be used as a storage medium for large amounts of data. If the SCSI adapters contained in the different computer systems use different mapping algorithms, the disk can not be read by another computer.

To resolve this shortcoming, manufacturers of SCSI-2 adapters often allow users to choose between two different BIOS drive systems, 4 or 8 Gigabyte. In many cases that does, however, require a *BIOS update* (Flash BIOS), whose implementation requires a certain basic knowledge on part of the user.

The same problem may occur when the SCSI adapter is exchanged in a computer, or a SCSI hard disk migrates from an old to a new computer. If a hard disk is not legible, although the SCSI bus is properly configured, you should check if the disk expects another mapping type than that used by the SCSI adapter.

3. 5 SCSI adapters and their installation

All theory is grey and problems lurk in the details. To be able to render a practical presentation of the simplicity or complexity of the installation of a SCSI adapter under different operating systems, and mention the problems that may occur in the configuration of the SCSI bus, I endeavour to present three SCSI adapters on the following pages which are distinctive not only as regards performance, but also price, while being typical of a certain market segment.

All three adapters have one aspect in common: they are connected to the PCI bus, thereby eliminating the risk of the computer bus becoming a bottleneck in the data transmission system.

General installation notes

No serious problems are anticipated when it comes to installing a SCSI adapter for the PCI bus. The computer and all peripherals connected to it are first disconnected from the mains, cable connections that are in the way are removed, and the computer case is opened. Most motherboards have three PCI and four ISA slots. Select a free PCI slot (PCI slots are white, in general, and shorter than the adjacent ISA slots), and then remove the slot bracket at the rear of the computer case.

Precautions must be taken against static electricity when fitting the adapter board into the PCI slot. The plug-in card contains CMOS components which are damaged or even destroyed when

3. SCSI in practice

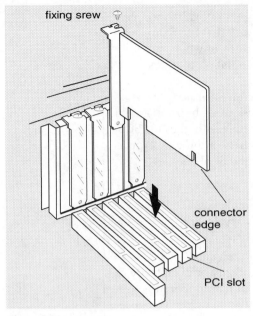

Figure 3.26.
Inserting a plug-in card into a PCI slot.

their gate inputs receive a voltage which is even a few millivolts higher than the supply voltage. Holding the card in your hands while inserting it, the supply voltage is, of course, 0 V. Even a small static charge at the gate inputs with respect to the supply voltage connections may be sufficient to damage components. Therefore, touch the metal parts of your computer to get rid of any static charges before touching the card. If it is necessary to put the card aside for a moment, it should be placed on its anti-static package.

The extension card is inserted into the bus slot, and the card mounting bracket is screwed to the rear panel (see Figure 3.26).

If the adapter may be terminated by software, you need not worry about the configuration of the SCSI bus as you fit the insertion card. If, on the other hand, the termination is switched on and off by a jumper, you may save yourself a lot of work and annoyance by considering beforehand whether the adapter has to be terminated or not.

If internal SCSI devices are to be connected, which is practically always the case unless you forgo SCSI hard disks, the internal flat-cable should be plugged into the corresponding connector on the adapter card. Do take care to get the polarisation right, because a small notch and a corresponding clearance may not always be available to guarantee the correct connection (Figure 3.27). In the absence of this mechanical aid, the pin-1 marks on cable (coloured cable side, small triangle marking on plug) must be made to match that on the connector (mark '1' or triangle). The same applies when this cable is used to make the connection with one or more SCSI devices inside the computer case (Figure 3.28). Normally, a SCSI adapter card comes with an internal connection cable having three

3.5 SCSI adapters and their installation

plugs. If more than two internal devices are to be connected, you need to get back to your supplier. External SCSI cables are, in general, not included with the SCSI adapter kit, and have to be purchased separately.

Notice:
The main part of this book discusses *Single-Ended Devices* only. Accordingly, only cables for this device type may be used.

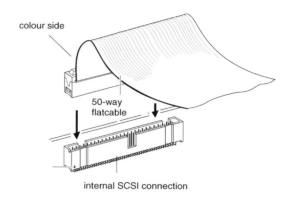

Figure 3.27.
When a flatcable is plugged on to an internal SCSI connector, the polarity should be observed.

PCI requirements

Two of the three SCSI adapters for the PCI bus function as a PCI master, that is, they bypass the CPU when it comes to data transfer to and from the main memory (*DMA = Direct Memory Access*). The operating system allowing, the CPU may handle other tasks in the mean time. This is only possible, however, if the adapter card sits in a *Bus Master compatible* PCI slot. Modern PCI board have this particular type of slot only. If you are not sure about your specific motherboard, the answer is probably in the *User Manual*.

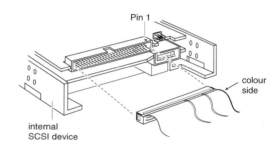

Figure 3.28.
Connection of the SCSI cable to an internal SCSI device.

Host adapters operating as PCI slave devices (data transfer via *programmed I/O, PIO*) exist by the grace of the DOS/Windows operating systems only. They do not allow bus master features to be used anyhow. For those of you considering a move to more up to

167

date operating systems like OS/2, Windows NT and LINUX, purchasing such PCI slave devices is out of the question in any case.

The parameters for the individual PCI slots are normally defined via the PC's CMOS setup utility. A number of points should be observed here:

- The available interrupts in a PC are divided between ISA and PCI slots. The PCI bus has four INT lines available: INTA# through INTD#. All INT lines are accessible on all PCI slots, and they are normally assigned IRQs by the BIOS. In this process, the number of IRQs distributed is the same as the number actually required by the insertion cards. In general, the host adapter expects the IRQ on the INTA# line of the PCI bus. If an additional INTA# setting is available in the BIOS setup, then this should be used.
- On the PCI bus, a distinction is made between *level* and *edge* triggering. With edge-triggered extension cards (this trigger mode was taken over from the ISA bus), the associated interrupt is normally defined with the aid of a jumper. Obviously, the setup may not be used in that case. The majority of PCI extension cards does, however, use level triggering, this is also applicable to the SCSI adapters discussed here. Consequently, the entry *Trigger Method* in the PCI setup must be set to the option *Level* for the PCI slot in which the SCSI adapter is fitted.
- The CMOS setup of some PCs allows PCI slots to be de-activated. If that is the case, you should make sure that the PCI slot for the SCSI adapter is *enabled*.
- To guarantee the correct allocation of free IRQs on to the PCI slots, the IRQs used by ISA plug-in cards must be marked as *used* with some boards.

Symbios 8250S

The SCSI adapter type 8250S from Symbios (formerly NCR) is an 8-bit adapter (PCI Bus Master) which, like most adapters in the 8xxxS family, is marked by an excellent price/performance ratio.

The board is equipped with the Symbios 53C825 processor, which establishes the link to the SCSI or PCI bus (32-bit DMA bus master), and generates the necessary timing. The adapter supports

3.5 SCSI adapters and their installation

Figure 3.29.
The PCI (Bus Master) to SCSI adapter type 8250S from NCR (Symbios), which is difficult to outclass in respect of price/performance ratio.

the SCSI-2 protocol (single-ended), allows synchronous and asynchronous data transfer, and may be actively terminated by means of jumpers on the board. The internal bus is implemented via the regular 50-way pinheader block, while the external connection is via the currently usual 50-pin HD socket. The TERMPWR line supply has an auto-resetting 1.5-A circuit breaker. The 40-MHz oscillator allows Fast-SCSI timing to be used, and a data transfer of up to 10 MByte/s to be achieved.

The adapter may be installed in parallel with any other hard disk controller (IDE, ESDI, ST506).

It is interesting to note in this respect that the Symbios 53C8xx chip is frequently used as the on-board SCSI adapter on computer boards of the ASUS company. In this way, this adapter type has gained wide acceptance and application. The on-board adapter then uses SDMS support (*SCSI Device Management Software*), which is contained in the Flash BIOS of the ASUS computer.

3. SCSI in practice

The Symbios 53C825 on our adapter card is actually a 16-bit Wide-SCSI processor, which is used in 8-bit mode here. As a result, a Wide-SCSI adapter is available under the name Symbios 8251S which exploits the full capacity of the processor. The use of a Wide-SCSI processor on an 8-bit adapter card is probably due to production methods — it is probably cheaper to use one and the same processor for different adapters, than produce a specific type for each card.

Drivers for DOS and Windows 3.x

The 100-odd pounds you have to fork out for an 8-bit SCSI adapter from the Symbios 8x50S series is an almost beyond competition price if you look at the performance offered. As may be expected, the cost of the kit also buys you a deluxe setup program. A simple installation routine which runs under DOS allows you to select between 'automatic' installation without the ability to correct, and a so-called 'custom' version that does allow the user to correct certain settings — simple and practical.

BIOS version 3.x

Two different CAM drivers, DOSCAM and MiniCAM, offering different levels of performance, a driver for hard disks and removable disks (SCSIDisk), a CD-ROM driver and a driver for the ASPI interface are available for use under DOS/Windows. Drivers for tape streamers or scanners are not found. Figure 3.30 once again clarifies the structure of the SDMS system from Symbios Logic with the CAM interface taken into account. The SCSI BIOS forms a fixed and indispensable part of the system (BIOS versions 3.04 and 3.06 allow hard disk capacities up to 4 Gigabyte, while version numbers 3.07 and up can handle sizes up to 8 Gigabyte, see *SCSI BIOS boot helper*).

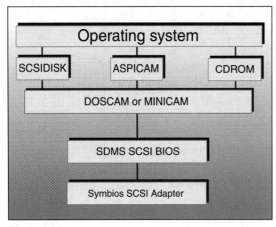

Figure 3.30.
Basic construction of the SDMS system and the driver supplied for the Symbios 8250S adapter card.

3.5 SCSI adapters and their installation

Working together, SDMS BIOS and Int13h enable up to eight hard disks (SCSI and others) to be connected under DOS (5.0 and later), without the use of additional drivers. The maximum number of SCSI drives is, of course, limited to seven because of the 8-bit wide databus.

It should be noted, however, that the 3.0x BIOS without CAM drivers offers limited VDS support only (*Virtual DMA Service*), so that the *DOSCAM* driver is required anyhow for fast DMA transfer.

A short aside: A Bus Master Device may address memory directly by bypassing the CPU or its DMA controller (with the exception of the AHA-2920, the PCI SCSI adapters discussed here are Bus Master devices).

Today's computers are generally based on 486 or Pentium processors which are capable of addressing so-called virtual memory. This may be done in *Virtual 8086 Mode* ever since the 386 processor was introduced. This mode allows the 386 processor and its follow-up types to extend the rather small 640-kByte DOS memory by means of *Memory Paging*, or emulate several 8086 processors, if necessary.

Unfortunately, Bus Master devices are incompatible with the *Virtual 8086 Mode*, since the memory is addressed directly with DMA transfers, neglecting any virtual memory. A solution for this problem is presented by the so-called VDS specification (*Virtual DMA Service*) which is supported by Microsoft, IBM and many hardware manufacturers. VDS-compatible drivers (and, with restrictions, SDMS BIOS 3.0x also; but version 4.0 to the full extent) allow Bus Master devices to find physically addressable memory even if the processor is operating in virtual mode.

CAM drivers

Remarkably, Symbios Logic supplies two different CAM drivers in spite of the limited range of drivers. *DOSCAM.SYS* is the complete, and *MINICAM.SYS*, the economy version. The fact that the economy version has right of existence becomes clear if you look at the size of DOSCAM.SYS — 41 kBytes is rather a lot which you would like to see stored in HMA memory to start with. By contrast, MINICAM.SYS weighs in at 10 kBytes only, a much handier format!

MINICAM.SYS

The function of *MINICAM.SYS* is to drive simple SCSI I/O functions and to support the ASPI interface in conjunction with ASPI-CAM.SYS.

MINICAM.SYS supports

- SCSI I/O functions (single thread);
- co-operation with several host adapters (in conjunction with SCSIDISK.SYS);
- logic sub-units (LUN).

DOSCAM.SYS

By contrast, DOSCAM.SYS may feature the following **extra** functions:

- capable of using synchronous data transfer (including Fast-SCSI);
- full VDS support (*Virtual DMA Services*);
- allows device control via *Tagged Queues*;
- multi-threading I/O functions;
- disconnect/reselect functions;
- Wide-SCSI data transfer*.

* Wide-SCSI may, of course, be realised with Wide-SCSI adapters from the Symbios 825xS family only. Adapters that do not have a Wide-SCSI connector are unable to operate in Wide-SCSI mode, not even if the processor chip and the CAM driver are designed to do so.

The term *multi-threading* describes a situation on the bus in which the SCSI adapter is processing commands addressed to two or more SCSI devices. In this mode, queue management becomes increasingly significant, too, because *multi-threading* and *tagged queues* in conjunction provide the Target with better opportunities for optimising the 'bus route plan'.

The category 'uninterrupted bus traffic' also includes the functions *Disconnect* and *Reselect* which enable a connection with the Target to be interrupted and re-established later (see *Reselect phase* in the theory section of this book).

3.5 SCSI adapters and their installation

Probably the most evident example of the need for a connectivity interruption between an Initiator and a Target is the rewind operation of a tape streamer. Rewinding a tape may take minutes, during which a number of other tasks may be carried out. The example becomes even more dramatic by imagining that a PCI slave (CPU-controlled) were active instead of our Bus Master adapter, and the PCI slave were unable to perform the Disconnect/Reselect functions — the result would be a total CPU blockage during the entire tape rewind operation.

Via the SDMS functions of SCSI BIOS 3.0x, SCSI hard disks may be addressed directly and without the use of additional drivers. CDS support for DMA transport is limited in this mode, which may cause considerably reduced performance particularly with fast hard disks (data transfer rates dropping by 20 to 40 per cent). On the other hand, no memory (HMA) is required for drivers, leaving precious space for other purposes.

Those of you who operate disk drives only on the SCSI bus should select the better alternative for themselves: the space-saving but slower attachment via the SCSI BIOS, or the faster data transfer by way of CAM drivers.

Any user is quite easily capable of gauging the difference in system performance with and without CAM drivers. Simply copy (under DOS!) a file of at least 10 MByte from one hard disk subdirectory to another. Use a stopwatch to measure the times needed for the copy operation, with and without the CAM driver installed. Depending on the difference between the two stopwatch readings, you may decide for yourself whether or not to install the CAM driver.

Device driver

You can not go round the use of a CAM driver if the computer is equipped with several SCSI adapters and/or SCSI equipment is employed which may only be addressed via device drivers (CD-ROM, removable disk drives, etc.). This is because device drivers may only be addressed via a software interface (CAM or ASPI), and that requires a CAM driver (MINICAM or DOSCAM) in any case, the BIOS being unable to provide further assistance at this level.

SCSIDISK.SYS

The device driver for magnetic disks offers control possibilities which exceed the functionality of the SCSI BIOS. The driver is required whenever magnetic disks are in use whose sector size differs from the usual 512 bytes. Moreover, it allows the operation of removable disk drives whose media may have to be removed/inserted with the computer running (that seems to be the case normally). The BIOS functions alone may suffice if these drives are used as hard disks, i.e., without media changing.

Additionally, if *SCSIDISK.SYS* is used, it becomes possible to operate more than eight disk drives under DOS, too (version 5.0 and up); one E-IDE type plus seven SCSI types (assuming an 8-bit SCSI adapter is used) or even more than seven SCSI drives (assuming Wide-SCSI is used).

The individual characteristics of *SCSIDISK.SYS* are:

- supports removable disk drives;
- allows the use of non-512-byte (standard) sector sizes (1,024; 2,048; 4,096);
- recognizes logic sub-units (LUNs);
- allows the use of several SCSI adapters (in connection with *DOS* or *MINICAM.SYS*);
- provides write protection functions;
- allows drive letters to be reserved

The last point becomes important especially when removable media having several partitions are involved. DOS normally reserves a drive letter ('station name') for any partition of a drive. If, however, no medium is installed in the removable drive when the computer is started, DOS reserves only one letter for this drive. Only the first partition is recognized if a medium having several partitions is inserted later.

To enable users to correct this situation, *SCSIDISK.SYS* offers the option /UNITS=x:y:x:z. If there is only one host adapter in the computer, the SCSI bus has number *0*, let the ID number of the removable disk drive be *4*, and let a removable medium have 3 partitions. The parameters after the SCSIDISK entry would then look like this: /UNITS=0:4:0:3. This statement reserves three drive letters for the removable disk drive.

The option /*Protect=[drive letter]* actuates the write protection for the indicated drive. Each write access attempt is answered by an error report.

The use of *SCSIDISK.SYS* (and all other device drivers) is only useful when a CAM driver was installed previously. Consequently, you should watch the order of the drivers in the config.sys file:

```
DEVICE=C:\[path]\DOSCAM.SYS /Option
DEVICE=C:\[path]\SCSIDISK.SYS /Option
.
.
.
```

In my experience, CAM drivers may be loaded into HMA memory, while SCSIDISK.SYS should remain in the main memory, 'high' loads often causing problems.

Since the introduction of Windows 3.11 (Windows for Work-Groups), it is known that there are 16-bit and 32-bit drivers under Windows and, consequently, 16-bit and 32-bit hard disk access operations. The standard drivers allow 16-bit access to be implemented. Unfortunately, there are no 32-bit drivers for BIOS version 3.0x of the SCSI adapter. From version 4.03 onward, such drivers are available (see below).

CDROM.SYS

Although the operation of CD-ROM drives on the SCSI bus invariably requires a device driver (and, of course, a CAM driver installed), the CD-ROM driver alone is not sufficient. DOS, and, consequently, Windows 3.11, requires a system extension that allows CD-ROM drives to be recognized. This extension is called *Microsoft CD-ROM Extension*, a.k.a. *MSCDEX.EXE*.

The abilities a CD-ROM drive is capable of demonstrating are dependent on the CD-ROM driver and its performance. Today, it is taken for granted that multi-session CDs can be read, although Video CDs or CD-I discs may not be processed by all driver/drive combinations.

If CD-ROM changers are to be driven, the driver should be capable of addressing logic sub-units (LUNs). For this purpose, CDROM.SYS offers the option /UPTOLUN=x, which allows the number of LUNs (0-7) to be defined. A drive letter is then reserved for each logic sub-unit. This option should only used if a CD-ROM changer is actually employed; the default indication for a regular CD-ROM drive is LUN=0.

The extension /D:[ident] is by no means optional. The ident is reserved for a code word with a maximum length of eight characters that functions as a transfer parameter. When MSCDEX.EXE is called in the AUTOEXEC.BAT file, this transfer parameter is also attached, establishing a clear and uniform relation between the operating system extension and a driver.

For example, if the entry in CONFIG.SYS reads

```
DEVICE=C:\[path]\CDROM.SYS /D:CD001
```

the call to *MSCDEX.EXE* should look like this:

```
C:[path]\MSCDEX /D:CD001
```

The transfer ident may be freely chosen, so *pipapo13* is also fine. The only thing to keep in mind is that the entries are identical.

Drive letter allocation

Under DOS, drive letters (stations) are allocated in a continuous order. The booting disk is allocated drive letter *C:*, then follow additional hard disks (if available) with the next letters in the alphabet, possibly followed by hard disk partitions and other SCSI devices which have been attached to the system by drivers. A CD-ROM drive receives its drive letter or letters (multiple LUNs) when *MSCDEX.EXE* is called. This usually means that the letter for the CD-ROM drive is the last in the series because all other drivers are called in *CONFIG.SYS*, while MSCDEX.EXE is not called until *AUTOEXEC.BAT* is run.

This last position in the series of drive letters does not present problems as long as the same letter is involved all the time. If, how-

3.5 SCSI adapters and their installation

ever, the connected devices include, say, an external removable disk drive which is used from time to time, small problems may occur.

If the external drive unit is not used, our CD-ROM drive moves forward one position (letter), being allocated letter *F:* instead of *G:*. All installed programs that want to access the CD-ROM player, however, will continue to do so under the letter *G:*, and find nothing there, resulting in an error report.

To prevent such hiccups, the option /L: [*drive letter*] in *MSCDEX.EXE* allows the CD-ROM drive to be assigned a fixed letter (obviously not *C:*, but any letter which is sufficiently far away from the those used for other drives in the system).

```
C:\[path]\MSCDEX /D:CD001 /L:G
```

Multiple CD-ROM drives

With hard disk drives, it is perfectly normal for different units to use the same device driver; SCSIDISK.SYS may even be employed above the adapter level. But what is the situation with CD-ROM drives, for which identification transfer between device driver and operating system is required?

In practice, no problems occur. A second CD-ROM drive is recognized without difficulties, and neatly included in the drive letter sequence after the drive letter assigned to the first drive.

ASPICAM.SYS

Together with a CAM driver, this driver forms the gateway to the ASPI interface. Using this auxiliary construction it is possible to employ drivers supplied by equipment manufacturers (which normally prefer the ASPI interface) in conjunction with adapters of the Symbios 8xxxS family.

The ASPICAM driver has to be loaded whenever an ASPI device driver or an ASPI utility is to be employed. This additional access appears to be essential, too, because the Symbios adapter kit does not include device drivers (for DOS and Windows) other than CAM drivers for hard disk drives and removable disk drives, as well as CD-ROM drives. And what about tape streamers, scanners, MO drives, etc.? Admittedly, magnetic disks and CD-ROM drives are the most frequently applied SCSI devices that have found the widest use, but that does not mean that the other devices may be labelled 'exotic', or unworthy of driver support.

If the driver set that comes with the Symbios 8250S adapter looks scanty, that has to be considered in a relativele way.

As explained on the theory pages of this book, widely different recording standards are used on tape streamers. Likewise, scanners come with a wide variety of features. On many units, the scan operation is started manually, while others require a SCSI command for this purpose. Seen from this perspective, it is more difficult for SCSI adapter manufacturers to provide drivers with full functional support for these device classes, than for equipment manufacturers, who pack drivers with their devices that make use of a defined software interface.

In this way it is simply bad luck (or the wrong strategy) that most equipment manufacturers prefer ASPI instead of the CAM interface favoured by NCR/Symbios Logic. As a result, users are forced to resort to such willy-nilly solutions as the ASPI driver being built on the CAM interface. Although ASPI.SYS is not memory-hungry at a size of 2 kBytes only, the solution is still not elegant.

The fact that NCR (now Symbios logic) did build on the CAM interface at all, although it was developed later than ASPI, is probably explained by the influence of John B. Lohmeyer. He was both chairman of the X3T9.x/X3T10.x Committee (i.e., the ANSI workgroup which finalizes the SCSI standard and has a strong steering effect) and a member of the Board of Directors of NCR. Now, CAM is the interface propagated by the SCSI Committee, while ASPI was developed by competitor Adaptec.

BIOS version 4.0x

Having taken over NCR, Symbios Logic seemed poised to correct this company strategy. The drivers for SYMBIOS 4.0 SDMS released in June 1996 use the ASPI interface and offer 32-bit support for Windows 3.x. In addition, a *Low Level Format Utility* called *AS-PIFMT.EXE* was added to the driver package, allowing SCSI magnetic disks to be re-formatted. The need to re-format magnetic disks when a different mapping is used was already discussed

3. 5 SCSI adapters and their installation

under *SCSI BIOS boot helper / INT13h*; the formatting utility for BIOS 3.0x is called *SCSIFMT.EXE*.

So much for the positive news.

Although the SCSI adapters of the 8xxx family have a so-called Flash BIOS that allows the user to perform an update operation using a BIOS file on diskette, it is doubtful whether such a procedure may be recommended to computer newcomers. In any case, this really useful technique should not be misused to force computer owners to perform updates and 'down'dates all the time.

SYMBIOS 4.03		SYMBIOS 3.0
ASPI8XX.SYS	–	DOSCAM.SYS
SYMDISK.SYS	–	SCSIDISK.SYS
SYMCD.SYS	–	CDROM.SYS
WIN8XX.SYS	–	–
–	–	ASPICAM.SYS

Table 3.5.
Comparision between the new ASPI driver for the SYMBIOS 4.0 SDMS and the CAM driver of the 3.0x BIOS used till now.

At the moment, only DOS and Windows drivers are available for BIOS version 4.0. If you wish to use other operating systems (OS/2, Netware, Unix), or even have these 'in parallel' on one and the same hard disk, and having defined, via the boot manager, which operating system to run in specific cases, you are forced to first modify the contents of your SCSI BIOS. This is necessary because the drivers for all other operating systems require BIOS version 3.0x as before. So, if you use only DOS and Windows 3.x on your PC, and do not wish to forfeit 32-bit disk access under Windows 3.11, you are allowed to update the SCSI BIOS of your 8xxxS adapter. In all other cases, you should postpone doing so until all drivers for all operating systems are capable of working with this BIOS version.

Fortunately, a potential buyer of a SYMBIOS 8250S or its relatives need not worry too much about all these BIOS complications. Symbios offers decent driver support via the Internet (as does the competition, Adaptec), so that new drivers for all supported operating systems are easy to obtain at any time.

3. SCSI in practice

The pages on the Symbios site are 'mirrored' at several locations that offer shorter loading times, for example, the server of the RWTH Aachen, Germany. The fact that several mirror sites exist also confirms that NCR/Symbios Logic adapters have found widespread use thanks to their excellent price/performance ratio.

Table 3.6.
Driver support for the Symbios 8xxxS family is available via the internet.

> Symbios-Logic FTP Server:
>
> ftp://ftp.symbios.com/pub/symchips/scsi/drivers/
>
> Mirror site at RWTH Aachen, Germany:
>
> ftp://ftp.dfv.rwth-aachen.de/pub/msdos/ncr/
>
> Readers who do not have Internet access may find the drivers (available at the time of printing this book) on the companion CD-ROM.

Table 3.7 provides and overview of the achievable transfer rates under DOS and Windows 3.11. The measurement results were published in *c't* computer magazine. The generally bad transfer rates under Windows are caused by the use of Windows-Interna and 16-bit Windows system access. Using specially modified 32-bit drivers, the transfer rate should double, roughly, under Windows 3.11.

Type	Manufacturer	Av. value DOS MByte/s	Peak value DOS MByte/s	Copy-Test W 3.11 MByte/s
AHA2940	Adaptec	6.442	9.171	1.183
KT946C	Buslogic	5.278	8.368	0.722
NCR8100S	Symbios Logic	6.336	9.120	1.163
NCR8150S	Symbios Logic	6.327	9.118	1.206
NCR8250S	Symbios Logic	6.338	9.122	1.220
Wide-SCSI				
AHA2940W	Adaptec	10.095	16.912	1.307
NCR8251S	Symbios Logic	10.252	17.077	1.316

Table 3.7.
Data transfer rates measured by c't magazine on 8xxxS family adapters as compared with the competition.

3.5 SCSI adapters and their installation

To guarantee the maximum data transfer rate with DMA operations, the *Double Buffer* option of SMARTDRV.EXE may not be activated (no relevant entry in CONFIG.SYS).

The measurement values do indicate, however, that the Symbios 8xxxS adapter need not hide itself to its more expensive competitors as far as performance is concerned.

Error causes
With the adapter installed properly, the CONFIG.SYS entries of the drivers provided with the right path markers and arranged in the right order (CAM before device drivers), and none of the configuration rules violated as regards the installation of SCSI devices (ID numbers, termination), the system should work without problems.

If not, the computer may even 'hang' as it attempts to start up, and a debugging round is in order. Reduce the error cause(s) as follows:

☞ is the SCSI BIOS recognized when the PC starts up?
(a report similar to that shown in Figure 3.31 appears on the screen)

If not, the error is probably caused by incorrect settings in the setup of the relevant SCSI slot.

```
NCR SDMS (TM) V3.0 PCI SCSI BIOS, PCI Rev. 2.0
Copyright 1993 NCR Corporation.

NCRPCI-3.04.00

ID 00    QUANTUM LIGHTNING 540S
ID 01    QUANTUM ELS170S
ID 02    QUANTUM FIREBALL1080S
```

Figure 3.31.
This type of report is shown when the SCSI BIOS is recognized as the PC starts up.

Errors may also be caused if you duplicate ID number allocations.
As a matter of course, external devices have to be switched on to enable them to be recognized by the BIOS.

☞ Are the connected SCSI devices recognized by the device drivers?
(a report similar to that shown in Figure 3.32 appears on the screen)

Figure 3.32.
If the device drivers recognize the connected SCSI devices, this, or a similar, message appears.

```
NCR SDMS (TM) V3.0 SCSI CAM Driver
Copyright 1993 NCR Corporation. All Rights Reserved.
DOSCAM-3.01.06
Board Count = 0001
PATH 0 is a V3.0 SDMS (TM) BIOS at 0004200
with IRQ = 09 DMA = 0
PATH 0, ID 0, LUN 0 is QUANTUM LIGHTNING 540S
PATH 0, ID 1, LUN 0 is QUANTUM ELS170S
```

If that is not the case, you should check to make sure that the CAM drivers are loaded before the device drivers in CONFIG.SYS. If you do not find errors here, and you are sure that the bus is properly configured (termination, ID numbers, etc.), the CAM driver may not work correctly with the memory manager (EMM386, QEMM, etc.). Try to load the CAM driver before the memory manager.

☞ When all these error causes may be excluded, and the system works, but not as it should, you should check if the TERMPWR line receives its supply voltage as it should. Multiple powering of the line may cause unexpected errors, while no voltage at all on the line is sure to bring the system to a halt.

Drivers for Windows 95

Driver installation is far smoother under *Windows 95*. Only one driver called *SYMC8XX.MPD* is required. Based on the requirements of Microsoft, it is built for so-called Miniport drivers, supports PCI/SCSI adapters, and allows the use of magnetic disks, CD-ROMs, tape streamers and devices from other classes. To enable a SCSI device to be addressed, device class drivers have to be implemented under Windows 95, too. These drivers are supplied by

3.5 SCSI adapters and their installation

either Microsoft or the relevant device manufacturer — *SYMC8XX.MPD* requires no modifications.

Using SCSI direct access and *SYMC8xx.MPD*, applications may talk directly to connected SCSI devices (*SCSI pass-through facility*, see the Windows 95 documentation).

The driver supports the following functions:

- synchronous data transfer (including Fast and Ultra-SCSI*);
- Wide-SCSI (when using an appropriate adapter);
- support for multiple host adapters;
- logic sub-units (LUNs);
- disconnect/reselect;
- extensive DMA transfer;
- SCAM function (**S**CSI **C**onfigured **A**uto**M**atically)*

Functions marked with an asterisk (*) are SCSI-3 precursors, discussed in detail in the sections *Wide-SCSI* adapters and *Plug & Play*.

Windows 95 also provides a built-in NCR810 driver, allowing simple adapters from the Symbios 8xxxS family to be used without having to install a new driver. This driver is particularly useful for the SC200 adapter which was sold in large quantities as a PCI/SCSI adapter for ASUS boards. Symbios SCSI adapters are recognized during the installation of Windows 95, and they are automatically installed provided the SYMC8XX.MPD driver is not required.

If the *SYM8CXX.MPD* driver is to be employed, different starting points should be taken into account during the installation:

- you are installing Windows 95 from scratch;

- you wish to update an existing installation of Windows 95, while
  a) no SCSI drivers were used so far,
  b) DOS drivers were used for the links to available SCSI devices.

Although Symbios Logic supplies extensive installation guidance for all situations (95.txt, which may also be found on the companion CD-ROM in the directory which contains Symbios drivers for Windows 95), you should be in the clear about the state of your system at the start of the installation.

System crashes during the installation, or error reports (yellow exclamation mark on the host adapter, or a red cross in the Device Manager) are generally not caused by Windows 95 or the new SCSI drivers, but by faulty (i.e, duplicate) IRQ allocations, a wrong DMA channel or another address conflict.

The statements about PCI setup made in the section *Drivers for DOS and Windows 3.x* are applicable to all operating systems, including Windows 95.

In case several Symbios adapters are integrated in the system, some fine-tuning may be required based on the fact that a host adapter is controlled by the NCR driver that comes with Windows, whereas the new driver supports all Symbios adapters in the system.

There are two ways of making sure that all Symbios adapters are controlled by one driver only.

1. Using the first method, all changes in the *Device Manager* are carried out by Windows 95. With each SCSI driver in the system, the used driver is replaced by the desired driver, and the concluding prompt for a new system start is answered with *No* until all modifications have been made.

In case the system is started from a hard disk, the order of the modifications should be fixed such that the driver for the adapter controlling the boot disk is changed last. After that, you may allow the system to restart.

2. Method 2 calls for a little more manual work. First, you find out which driver is used for all adapters by looking in the resources allocation overview of the Device Manager. Go to the subdirectory \WINDOWS\SYSTEM\IOSUBSYS, and temporarily rename the relevant driver (use a different name than *.MPD). After a system restart, Windows may want to post-load the Bundle driver. Confirm with *OK*. Next, if a message appears telling you that a driver

3.5 SCSI adapters and their installation

could not be found, click *Cancel*. All further messages concerning SCSI adapters are answered in the same way.

If the boot-up operation is finished, change the driver/adapter allocation in the Device manager as described with Method 1 (if the desired driver is the one that was renamed earlier, this change has to be undone). Do not allow the system to restart until you have made all changes.

The SYMC8XX.MPD driver on the companion CD-ROM is version 2.0 already, in which a number of teething problems were solved. It now supports chip types 810A, 825A, 860 and 875.

Drivers for OS/2

Under OS/2, too, SCSI firmware is capable of addressing connected SCSI hard disks directly through the BIOS. To improve the system performance, however, it is recommended to install a driver called *OS2CAM.ADD*. As with Windows 95, it is the only driver required to compete the operating system.

OS2CAM.ADD extends the SCSI system with the following functions:

- synchronous data transfer (including Fast-SCSI);
- Wide-SCSI transfer (single-ended and differential when using suitable adapters);
- multiple host adapter management;
- disconnect/reselect functions;
- full DMA support;
- tagged queue control.

The driver complies with the IBM OS/2-ADD Specification (*ADD = Adapter Device Driver*) and is, of course, also required for all SCSI devices that can not make use of direct BIOS support (CD-ROM drives, tape streamers, scanners, etc.). The device drivers required for this function are either contained in the OS/2 package (standard devices), or supplied with the relevant SCSI equipment. The device drivers have to be included in the *CONFIG.SYS* file (see OS/2 manual). You have to mind the order of the driver entries, because it is used by OS/2 to allocate drive letters.

3. SCSI in practice

The installation of the *OS2CAM.ADD* driver should not present difficulties, the installation guidelines from Symbios and IBM helping you on. Successful installation is indicated by the line

```
BASEDEV=OS2CAM.ADD /V
```

in the CONFIG.SYS file. The driver provides several options (/V, /SN, /Q, /DM, /SM) that allow certain indications, the use of SCSI and DASD managers, but also synchronous data transfer and the use of LUNs to be switched on and off manually.

Drivers for Windows NT (3.51)

Under Windows NT, I/O operations are completed via the I/O manager. To enable a SCSI device to be addressed, the steps shown in Figure 3.33 have to be completed. Both the port driver and the essential device class drivers are included in the Windows NT package. Drivers for more exotic devices should be supplied with the relevant equipment, and may be installed without problems, and without the need for modifications to the Miniport driver. As a matter of course that also applies to standard devices, in as far as the driver supplied by the manufacturer offers more extensive performance. Decisive information may be obtained in this respect by comparing the dates of the drivers. Tape streamer support is integrated in the operating system, so that an additional class driver is not required.

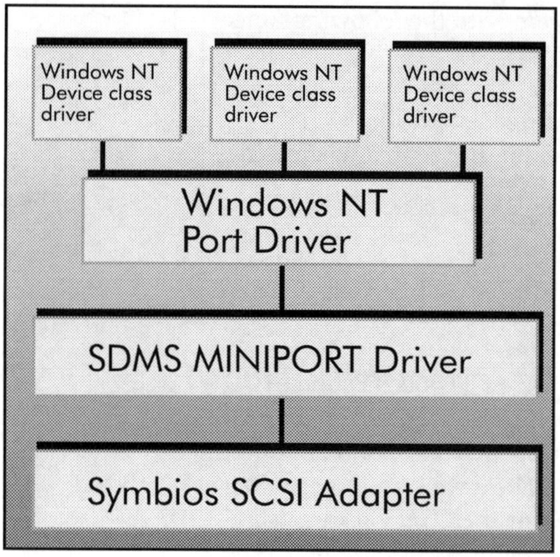

For the SDMS BIOS 3.0, NCR/Symbios supply a Miniport driver, NCRSDMS.SYS, which acts as a bridge between NT Miniport drivers and the SCSI chip, offering the following functionality:

Figure 3.33.
Individual levels of the connection between operating system and SCSI adapter under Windows NT.

3.5 SCSI adapters and their installation

- synchronous data transfer including Fast-SCSI;
- tagged queue control;
- multiple host adapter management;
- multiple LUN management;
- disconnect/reselect functions;
- full DMA support;
- Wide-SCSI support, provided the relevant adapters are connected;
- SCSI direct access support (Pass-Through Functionality)

The *Pass-Through Functionality* allows NT applications to have direct access to a SCSI device. Such possibilities are sadly lacking from the Windows versions that run on top of DOS.

NCRSDMS.SYS has no manual settings whatsoever, and its installation is smooth. None the less, if problems occur, these are generally caused by an incorrectly configured SCSI bus, or errors in the PCI setup.

Drivers for Netware 3.1x and 4.x

Since the change from Netware 3.1 to 4.0 is not, stricty speaking, an upgrade but more of a system change, these different family members require different SCSI drivers. This applies to the SDMS drivers as well as to the device drivers and the ASPI add-ons.

The drivers for the different versions do, however, offer the same functionality:

- ASPI interface support (using an add-on driver);
- synchronous data transfer (including Fast-SCSI);
- multiple host adapter management;
- logic sub-units (LUNs);
- disconnect/reselect functions;
- full DMA support;
- tagged queues;
- interrupt sharing;
- Wide-SCSI support using appropriate SCSI adapters;
- possibility to exclude individual devices.

3. SCSI in practice

The installation requires some manual work, though the installation guidelines are so extensive that it is not necessary to go into details here.

A number of optional extensions are available that enable functions to be switched on and off, and, more importantly, maximum values to be set up. These values have greater significance in a larger network than in the case of a stand-alone operating system.

- The option *tags = [disable/enable]* disables or enables operation with tagged queues.
- The option *depth = [0-128]* fixes the allowable depth of the queues.
- Using *timeout = [15 - 999999999]* a timer may be set that interrupts a certain action after a predetermined period [s].
- *sharedint = [enable/disable]* controls *Interrupt Sharing*. This function is needed when several SCSI adapters of different makes are operated on the SCSI bus. Any non-Symbios Logic adapter(s) must be able to handle interrupt sharing.
- *wide = [enable/disable]* allows manual control over the Wide-SCSI mode.
- *xcl = [path, ID, LUN : path, ID ...]* excludes individual devices or their LUNs. The indication is always by means of numbers in the order path, SCSI-ID, LUN. The colon (:) delimiter is used to separate the indications for two different devices.
- *max_HBAS = [1 - 8]* limits the number of SCSI adapters to a maximum value.
- *max_id = [8 - 32]* limits the number of allowable SCSI IDs to a predefined value.
- *max_lun = [1 - 4]* allows only a fixed number of logic sub-units.

The memory requirement of the driver depends on the number of host adapters, the allocated LUNs and SCSI IDs. It may be computed from the equation

```
Mem. requirement = 319  max_hbas  max_lun  max_id
```

The value is subject to large differences which are caused by the requirements (maximum value: 330 kBytes with 32-bit Wide-SCSI, eight host adapters and four LUNs per ID; minimum value:

3.5 SCSI adapters and their installation

2.5 kBytes with one host adapter, one allowable LUN and a bus width of eight bits).

Drivers for UNIX

With Unix drivers, Symbios Logic makes a distinction between drivers for UnixWare (1.x and 2.x0 and drivers for SCO Unix, the first operating system for IBM compatible computers). In both cases, a basic working knowledge of Unix is assumed when the installation is involved.

The installation help supplied by Symbios Logic (INSTALL.TXT in the respective Unix subdirectory on the companion CD-ROM) are so extensive that there is no need for me to discuss the integration of the drivers into the operating system in greater detail. I will therefore limit myself to presenting a short overview of the functionality of the adapter in conjunction with the relevant drivers.

SCO UNIX
- Provides SCSI configuration assistance
- Multiple processor support
- Synchronous data transfer (including Fast-SCSI)
- Wide-SCSI support (single-ended and differential)
- Disconnect/Reselect functions
- Full DMA support
- Tagged queues
- Supports dynamic interrupt mapping
- Shared interrupts are allowed
- Logic sub-units (LUNs) are recognized
- RAID functions up to block size of 1,024 bytes

UnixWare 1.x/1.2
- PDI-ID installation assistance
- Multiple processor support
- Synchronous data transfer (including Fast-SCSI)
- Multiple Host adapter management
- Disconnect/Reselect functions
- Logic sub-units (LUNs) are recognized
- Full DMA support
- Wide-SCSI transfer (version 1.1 is unable to recognize devices with an ID number > 7)

3. SCSI in practice

- Direct SCSI access (Pass-Through Functions)
- Tagged queues
- Supports request chaining

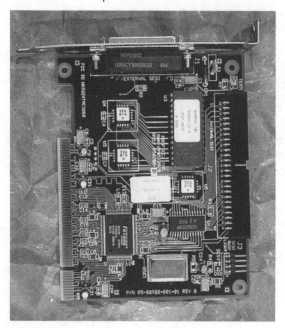

Figure 3.34.

The AHA-2920 – a SCSI adapter for the DOS/Windows environment (PCI Slave).

Adaptec AHA-2920

"Host Adapter AHA-2920 exploits the computing power of modern Pentium processors to achieve high data transfer rates"

Though pretty effective from an advertising point of view, this description is indicative of the fact that the AHA-2920 is a PCI slave, not a Bus Master adapter. It operates in *Programmed I/O Mode* and requires the performance of the CPU to achieve high data transfer rates, irrespective of the operating system it is used with (see *PCI requirements*).

Under DOS and the Windows versions that build on it, such an adapter is by no means a bad performer. After all, the operating system is incapable anyway of freeing the CPU when an I/O operation is being processed. As already mentioned, DMA is unable to exploit its power in this case.

The situation looks totally different under operating systems like Windows NT, OS/2, Unix and others. A SCSI adapter that has to wait for 'orders from higher authorities' (being a PCI slave), clearly takes a back seat to Bus Master adapters in respect of performance.

The CPU has to deal with the data transfer itself, which runs via *Programmed I/O (PIO)*, in other words, 32-bit ports, on the AHA-2920. As a result, the CPU is blocked until the end of the transfer,

3.5 SCSI adapters and their installation

and unable to carry out other functions. Though such a condition may be tolerated on a stand-alone computer, it is unacceptable on a server system.

Costing about £95, the AHA-2920 is an 8-bit PCI bus adapter which belongs in the low-cost range. That does not, however, explain its weakness in respect of the achievable data transfer rates, the Symbios 8250S costing about £99. On the positive side, the largely automatic operating installation software and termination checking functions offered by the Adaptec card make it interesting for SCSI newcomers.

Looking at the price and performance, the AHA-2920 is clearly aimed at desktop computers with a corresponding operating system. The kit includes an extensive software package that makes the installation a piece of cake while offering test utilities for the SCSI bus and bonus programs for tape streamers (backup software), and, importantly, CD-writers.

EZ-SCSI 4.0

The installation software called *EZ-SCSI 4.0* is not tailored to the AHA-2920 card, but capable of recognizing a wide variety of SCSI adapters from the Adaptec house. It does, however, operate in conjunction with Microsoft operating systems only (Windows NT, 95 and 3.1x).

That does not mean that the AHA-2920 (or other adapters) may not be installed under other operating systems like DOS, OS/2 and Netware — only the large number of enclosed utilities and test programs run on Windows platforms only. Whereas EZ-SCSI 4.0 fills an entire CD-ROM, the other installation routines come in much simpler apparel on a single diskette.

Although the DOS installation routine also goes by the name of EZ-SCSI 4.0, it has no utilities whatsoever, because Adaptec, leader in the development of the ASPI interface, has produced an ASPI for Windows in 16- and 32-bit versions only. None of the important auxiliary and test programs run on a Windows platform, but also require the special ASPI version. This limits the application of these programs to Windows and the SCSI adapters from Adaptec.

3. SCSI in practice

Figure 3.35.
Installation software EZ-SCSI 4.0 from Adaptec.

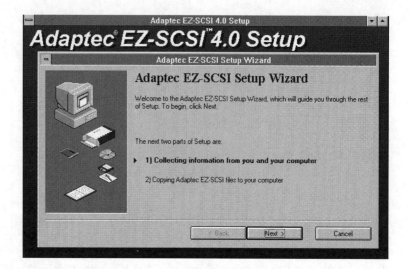

Adaptec wants to emphasize the performance of the AHA-2920 by listing the *Advanced Features:*

- 😐 10-MByte synchronous Fast-SCSI transfer
- ☹ Up to seven devices are supported simultaneously (*depending on the number of datalines on the SCSI bus, and not on the manufacturer's performance!*)
- 😊 **Plug and Play support** for hard disks, CD-ROMs, removables, scanners, tape streamers, Zip drives, MO drives, DAT recorders and other SCSI devices (*see section on Plug and Play*).
- ☹ Programmed I/O data transfer.
- 😐 Support for the following operating systems: DOS; Windows 3.1x, 95 and NT; OS/2 Warp; Novell Netware (*Unix drivers were not included with the adapter, although that is stated on the box*).
- 😐 operates with the 133-MHz fast PCI bus.

The Plug & Play functions deserve to be highlighted, all other '*Advanced Features*' are self-evident for 8-bit PCI/SCSI adapters. The PIO mode is noted as a negative aspect.

3.5 SCSI adapters and their installation

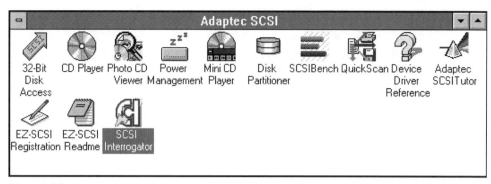

Figure 3.36.
Auxiliary and extension programs provided by the EZ-SCSI 4.0 bundle. These programs run under Windows only (3.1, 95 and NT).

The installation

Concerning the insertion of the card into a PCI slot in the computer, the same applies, basically, as what has been said about fitting the Symbios 8250:

- Take appropriate measures against static charges;
- The adapter expects a free interrupt on the INTA# line, and 'level' triggering;
- Observe the polarization of the plug terminating the 50-way internal SCSI flatcable as it is inserted into the mating socket on the adapter board (pin 1 goes to pin 1).

Connecting internal SCSI devices calls for a flatcable having three plugs (to connect to two devices), which is included with the kit. Internal and external devices are connected up (the latter using external round cabling, not included in the kit), and a terminator is fitted on the last device of each series. You need not bother about the termination of the adapter, because the card runs a check on the bus as the computer starts up, and performs the termination functions automatically.

Drivers are not necessary if SCSI hard disk drives are used only, the AHA-2920 having its own SCSI BIOS (an otherwise identical version without a BIOS is available as type number AHA-2950).

3. SCSI in practice

The driver installation boils down to "*start EZ-SCSI 4.0 and follow the directions*"; the installation program is really a piece of cake. A distinction is made between a *Typical* and a *Custom* installation. The latter allows the user to intervene at certain points, while also providing an indication of what is happening in detail. The program recognizes the Adaptec insertion card, installs the requisite drivers, and completes the ASPI manager with the necessary settings.

The most important options for the ASPI manager for DOS and Windows 3.1x are:

/1- Logic sub-units (LUNs) are not supported.
/u- Forces asynchronous transfer.
/y Switches off parity checking.

The case becomes a little more interesting when a totally new installation is involved, and the CD-ROM drive is not yet recognized because of missing drivers. There's no avoiding the diskette version in that case, to provide the elementary outfit in the first place.

The final clever bits, however, may not be configured automatically by EZ-SCSI — for example, if you want to reserve more than one drive letter for a removable disk drive, you may do this in menu-driven fashion under Windows 95 (function provided by the operating system). Under DOS and Windows 3.11, however, that requires manual work on part of the user. The entry for ASPI-DISK.SYS in CONFIG.SYS is completed by the option /r*x*, where *x* stands for the number of letters.

Figure 3.37.
Icon for the installation of 32-bit hard disk access under Windows 3.1x.

Adaptec was the first manufacturer to supply a driver with its SCSI adapters that supports the faster 32-bit hard disk access mode under Windows 3.1x. Normally, the driver would be installed via the *32-bit Disk Access* button, and a menu hiding below it. That, however, applies to Adaptec's Bus Master adapters, which adds another negative aspect to the AHA-2920's score.

The icon *32-bit Disk Access* does, however, offer another, additional, function which is quite useful. The write cache for SCSI disks may be switched on and off in the *Control Panel*. However, as for all utilities from the EZ-SCSI bundle, that only applies to devices controlled via SCSI adapters from Adaptec.

3.5 SCSI adapters and their installation

EZ-SCSI represents great installation software which particularly lightens the newcomer's introduction to SCSI. The extensive utilities and add-on programs (different numbers depending on the Windows system you are using) complete the picture. In addition to the add-on programs, which include jewels like the programs for copying and writing CD-Rs, and the QuickScan Tool, the SCSI auxiliary programs in particular should meet with great interest.

The *Power Management Tool* enables a timeout value to be determined after which the rotation speed of non-used hard disks is reduced. If necessary, the disks may be activated immediately again.

The *SCSI Interrogator* provides an accurate graphic overview of the configured SCSI system, providing *Lock* and *Unlock* commands for all removable media.

The *SCSI Bench* software and the monitor program *Drive Light* enable the performance of the system to be measured and the dates of use to be recorded.

A successful utility collection, which comes at a price, however. Depending on the adapter type, the software and the SCSI adapter are sold separately or as a kit. The AHA-2920 is only available bundled with EZ-SCSI 4.0. Other adapters, too, are equipped with this software, so that you are not forced to buy the AHA-2920 to obtain the installation program and the utilities collection.

Wide-SCSI adapters

As you may know, there are two ways to increase the data transfer rate on the SCSI bus — either you shorten the timing (shorter signal and delay times), or you widen the lane, that is, put more datalines at the system's disposal. The first option is called Fast-SCSI, the second, Wide-SCSI, and both acceleration methods may, of course, be used together.

Fast- and Wide-SCSI are negotiated between Initiator and Target in the Message phase — *"I prefer double lanes and top speed! What about you, or do you use country roads?"*

3. SCSI in practice

The points to note in practice when it comes to applying Fast- and Wide-SCSI are briefly listed below.

Using Fast-SCSI timing, which may, of course, be employed on the 8-bit wide bus as well, results in two differences for the user:

- the maximum allowable cable length is reduced from six to three metres;

- the SCSI bus should have active termination to guarantee clean signals and, as a result, error-free data transmissions.

Nearly SCSI compliant devices and those capable of handling the normal timing only may be used on the same bus, because Fast-SCSI is negotiated between Target and Initiator during the Message phase, as are synchronous and asynchronous data transfer. The only condition is that the SCSI adapter can manage Fast-SCSI.

Figure 3.38. Corruption of a rectangular signal by noise.

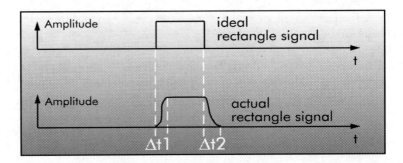

With the use of Wide-SCSI devices, the difference is clearer. On the insertion card of a Wide-SCSI adapter, a 68-pin HD connection has been added to the internal 50-way boxheader or pinheader block. The external Wide-SCSI link uses a 68-pin socket instead of the 50-pin HD connector.

3.5 SCSI adapters and their installation

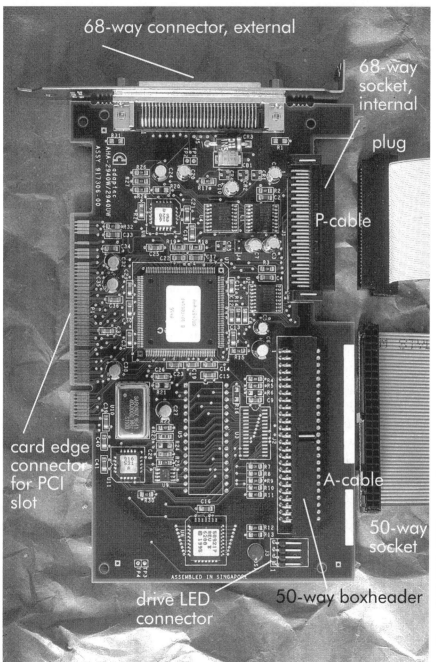

Figure 3.39. Wide-SCSI connections.

The resulting data transfer rates are roughly as follows:

- Asynchronous data transfer — max. 3.5 Mbyte/s;
- Synchronous data transfer — max. 5 Mbyte/s;
- Fast-SCSI timing with synchronous data transfer — max. 10 Mbyte/s;
- 16-bit Wide-SCSI using Fast-SCSI timing and synchronous data transfer — max. 20 Mbyte/s.

According to the SCSI protocol, synchronous data transfer, Fast-SCSI timing and the doubled number of datalines need not be linked to one another. So, asynchronous data transfer using normal timing values and a 16-bit SCSI bus is perfectly possible, although the combination is rarely seen in practice.

The increase in performance is always implemented using the smallest possible effort. To begin with, synchronous data transmission is employed, and then Fast-SCSI **with** synchronous data transmission. If the achievable data rate is still not high enough, you have to widen the bus **in addition** — Wide SCSI.

Consequently, you will look in vain for SCSI devices sporting a Wide-SCSI connection, but unable to handle Fast-SCSI timing. Similarly, Fast-SCSI devices are few and far between that do not transmit data in synchronous mode across the bus.

Comparing figures

To obtain exact indications on the actually achievable **net** data transfer, the control overhead has to be subtracted from the maximum data transfer rates mentioned above. The control overhead is the total amount of data required for controlling and maintaining all communication on the SCSI bus. If the overhead was relatively large under SCSI-1, the situation has improved so much under SCSI-2 that the maximum values may be considered actually achievable data transfer rates, at least for an initial approximation.

Comparing these figures with the transfer rates that can be achieved by SCSI devices may cause some raised eyebrows — only SCSI hard disk from the above-average class and better may appear to put some sense in using Fast-SCSI at more than 5 Mbyte/s. All other devices may, generally speaking, be satisfied with asynchron-

3.5 SCSI adapters and their installation

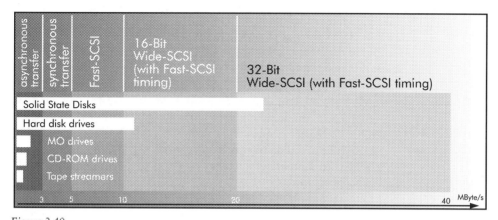

Figure 3.40.
Maximum data transfer rates that may be achieved by devices from the different device classes, in relation to Fast-SCSI and Wide-SCSI.

ous data transfer. Figure 3.40 indicates that the 3-MByte/s limit is not exceeded by devices not belonging in the *hard disk* class. True, some expensive hard disks like the Seagate Barracuda 2HP having two heads for each disk side are rare birds whose transmission rate of 10 Mbyte/s justify a Wide-SCSI interface. No other hard disk is slowed down, however, by an 8-bit SCSI system.

Consequently, SCSI devices having a Wide-SCSI connection are only found among hard disks. No Wide-SCSI devices exist outside the class of *Direct Access Devices*. In this area, however, a growing number of hard disks having transfer rates of around 7 Mbyte/s is fitted with a Wide-SCSI interface. For what purpose?

The answer is relatively simple. All you have to do is imagine that the DOS operating system with the Windows add-on is replaced by a multi-tasking and multi-threading system capable of exploiting the *Disconnect/Reselect* SCSI functions (Windows NT?), and that several hard disks are humming in computers running such an operating system. These disks would then be able to respond simultaneously to read and write requests, and transmit their data in parallel across the bus (both SCSI and the operating system allow that). The maximum data throughput of a hard disk (6-7 Mbyte/s) should then be multiplied by the number of available hard disks — with just two hard disks installed in the system, the Fast-SCSI limit of 10 Mbyte would be exceeded.

Summarizing: if you use a fast hard disk only, a Wide-SCSI interface is practically useless. Even if a cache memory with a size of 2 Mbyte is used, time savings are of the order of tenths of a seconds when comparing an 8-bit SCSI with Wide-SCSI interfaces. As soon as the cache has been read or written, (a 2-MByte cache is read in about 0.2 s at a data transmission rate of 10 Mbyte/s), access to the disk is direct again, which takes you back to the accustomed transfer rates again.

If the PC contains several hard disks, and the operating system allows parallel access, a worst case analysis shows that an 8-bit SCSI bus using fast-SCSI timing becomes a bottleneck when only two disks are accessed. In a computer matching this description, Wide-SCSI is sure to result in noticeable improvements as regards system performance.

32-bit Wide and Ultra-SCSI

If you keep on calculating you will find out that the 20 Mbyte limit may be broken using four Wide-SCSI hard disks. In such a case, another extension is required — faster timing or a wider bus.

As discussed in the theory chapter, a 32-bit wide version of the SCSI bus has been defined in addition to 16-bit Wide-SCSI. The 32-bit version allows the 20 Mbyte/s limit of 16-bit Wide-SCSI to be moved up to 40 Mbyte/s. In practice, however, 32-bit Wide-SCSI is hardly used, a 110-pin plug not fitting very well in the trend towards miniaturization.

Instead of it, *Ultra-SCSI* or *Fast-20-SCSI* is emerging as a quasi-forerunner of SCSI-3. Under Ultra-SCSI, the timing values are halved once again relative to Fast-SCSI, allowing a maximum speed of 20 MByte/s to be reached even on an 8-bit bus, or 40 Mbyte/s on a 16-bit Wide-SCSI bus (which is then referred to as a *Fast-40-SCSI* bus). With 32-bit Wide-SCSI, the transfer rate would be 80 Mbyte/s.

All of this is only possible if active termination is applied as well as cables of superb quality. Also, the cable length is halved once again, leaving you with a maximum length of just 1.5 m.

3.5 SCSI adapters and their installation

So-called *UltraWide-SCSI* adapters with a 16-bit wide bus and Fast-20 timing are already around, although I am not yet aware of any matching devices.

ID numbers

The advantage of a Wide-SCSI adapter is not just the higher data transfer rate, but also the possibility to connect up to 15 SCSI devices. Whether or not this option will make it to application in home PCs remains to be seen. For server systems, on the other hand, it may be of great interest.

Because a Wide-SCSI adapter allows devices having an 8-bit interface as well as real Wide-SCSI devices to be connected, you have to keep a few things in mind when allocating the ID numbers. As usual, the boot disk gets ID 0, while ID 7 is reserved for the adapter. All 8-bit devices then have to be assigned numbers in the range from 1 to 6. Consequently, the Wide-SCSI devices should get numbers > 7, because 8-bit devices allow IDs between 0 and 7 to be set only.

Termination

Because at least 8-bit devices (50-way connector) and 16-bit devices (68-way connector) may be hooked up to the internal side of a Wide-SCSI adapter, and you can not uphold that one device follows the other, i.e., a chain is formed, the question arises what the termination of such a system should look like.

Let's have another look at Figure 3.3 to make answering this question a little easier. As you can see, each individual signal wire has to be terminated. The terminators discussed so far (active or passive) for the 8-bit bus contain a resistor network that keeps each individual line at a nominal potential.

The SCSI bus is 16-bit wide now and taken out on two connectors. To fit each signal wire with a wire termination, each of the two connectors, or the last device connected to it, has to be terminated. If 8-bit and 16-bit devices are operated on the internal connection, each end device has to be terminated on the respective cable (50-way or 68-way). By contrast, if a cable is not used (only 8-bit and 16-bit devices are connected), the non-used cable must be terminated at the adapter.

3. SCSI in practice

At the external side, things are simpler or more complex, depending on how you look at the situation. The external connection with 16-bit Wide-SCSI consists of a 68-way HD socket which accepts a direct connection to Wide-SCSI devices only. The last device at the external side is terminated as usual, and problems with a second line do not occur. This is the simplest solution.

The situation becomes more complex if 8-bit devices are to be used on an external Wide-SCSI connection. This requires an adapter which reduces the 68 Wide-SCSI connections to 50. In the adapter, the non-used wires may not be left open-circuited because they carry signals that are responsible for Wide-SCSI. These lines should therefore be terminated (actively). Such an adapter is supplied by, for example, Adaptec as type number *ACK68P-50P-E*. It features a 68-pin HD plug and a 50-way HD socket.

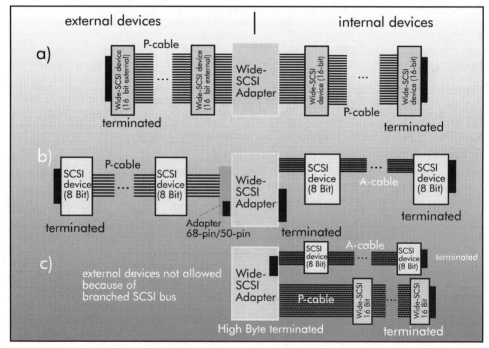

Figure 3.41.
Termination of a Wide-SCSI system in case (a) only 16-bit devices or (b) only 8-bit devices or (c) devices for both bus widths are connected.

3.5 SCSI adapters and their installation

This combination of an adapter and a terminator is simply plugged the regular external Wide-SCSI connection, while the other side links to a 50-pin HD plug as normally used for 8-bit connections. The relations in the termination of a Wide-SCSI bus are clarified once again in Figure 3.41.

Meaningful distribution of data

To maintain a possibly high data throughput on the Wide-SCSI bus when several hard disks are employed, the user may lend a helping hand, but only if the operating system allows parallel access to different devices (Disconnect/Reselect functions).

The aim is as follows: files that are frequently used should be distributed across as many disks as possible in order to increase the probability of simultaneous access to different data carriers. This aim is relatively simple to achieve when storing a swap file for the operating system. As a rule, the swap files should be stored safely on a disk which is not so often used, so that storage of files with memory-intensive applications and access to new files may be run in parallel.

Windows NT has another option available that allows several hard disks to be bundled into a so-called *Stripe Set*. In this system, files to be written are subdivided into logic segments which are distributed across different physical disks. Segment 1, for example, goes to disk 1; segment 2 to disk 2; segment 3 to disk 3; segment 4 to disk 1 again, segment 5 to disk 2, and so on. In this way, large files are distributed across several hard disks, enabling several hard disk access operations to take place when these files are read. The result is a considerably higher data throughput on the Wide-SCSI bus.

On the down side, data security is reduced by this method. If, for example, one hard disk breaks down in the above 3-disk system, not one third of the files is lost, but one third of each file, which is the same as total data loss.

RAID systems

A good balance between data throughput and data security is offered by the different levels of so-called RAID systems (*Redundant Array of Inexpensive Disks*). A distinction is made between levels 0 through 6, where

RAID 0 is nothing but a file distribution method much like that of *Stripe Sets*.

RAID 1, also known as *Mirroring*, is a system in which written data has duplicate presence on two different data carriers. While data security is increased, data throughput drops.

RAID 2 employs the disks as with RAID 0, although at least one disk is used as a so-called Check-Disk for error correction. The advantage of this method is mainly higher data throughput when reading large files, because their contents have been distributed across different disks as under RAID 0, where the error probability is clearly smaller than with RAID 0 thanks to the Check-Disk.

The disadvantage of this method becomes apparent when files are being written. Because an error correction entry has be left behind on the Check-Disk for each file, this disk has to be accessed with each write operation, and that may cause a bottleneck. In any case, the write transfer rate is smaller than the data throughput achieved with read operations.

RAID 3-6 makes about 80 per cent of the total disk capacity available for data storage, using the remaining 20 per cent for the storage of parity checking codes. With RAID 3 and 4, these codes are stored on an individual disk (slowing down data throughput with reading and writing). With RAID 5, storage capacity on several disks is used for this, while RAID 6 always attaches the parity information of the previous segment to the next data block, which is stored on a different disk.

To enable such security measures in the distribution of data to be implemented, special SCSI adapters are available having an additional co-processor which is in control of the RAID functions. In general, such an adapter is capable of distributing the hard disks across three independent 8-bit buses. The *AHA-3985* from Adaptec is such a SCSI adapter offering RAID functions.

3.5 SCSI adapters and their installation

If Wide-SCSI adapters are to be integrated into such a system, the operating system must take over the control of the RAID functions. Up to now, that is only possible under Windows NT by using the *Stripe Sets* (RAID 0) mentioned earlier. For Linux, a driver (Beta version) is available which should take over similar functions.

Disconnect/Reselect with hard disks

Because data distribution across several hard disks as described here only enables a performance improvement to be achieved if the operating system allows DMA transfer as well as *Disconnect/Reselect* functions, it may be useful to elucidate the timing of the break and reconnect instants of a link, using the operations used for reading a hard disk.

In the following example, the # character indicates a break in the connection between Initiator and Target, while → stands for reconnection.

The hard disk receives a read request.
#
It positions the read head over the desired cylinder and reads a certain amount of data (defined via SCSI Modepage) into the data buffer.
→
It copies the data to the Initiator.
#
The data buffer is filled again.
→
Data are transferred to the Initiator.
#...→...#
Once all requested data have been transferred, the connection is finally broken.

During the periods when the connection between the hard disk and the Initiator is broken, the Initiator is allowed to arrange other data transmissions.

3. SCSI in practice

The conditions for fast data transfer under Wide-SCSI to exceed what may be achieved by an 8-bit bus are, therefore:

- several possibly fast hard disks must be used, across which the files may be distributed in a meaningful way, and
- an operating system must be used which supports DMA transfers, and is capable of employing the *Disconnect* and *Reselect* SCSI functions.

Using a Wide-SCSI adapter under DOS or a similarly dated system is a complete waste of money.

AHA-2940 UW

Future-oriented, but at the right time — the AHA-2940 UW from Adaptec is a 16-bit Wide-SCSI adapter with fast-20-SCSI timing, in other words, an Ultra-Wide adapter.

Figure 3.42.
The AHA-2940 UW – 16-bit Wide-SCSI using Fast-20 timing.

3.5 SCSI adapters and their installation

The main features of this adapter are:

- 40 Mbyte/s maximum data transfer rate at 16-bit bus width and Fast-20-SCSI timing;
- 20 Mbyte/s when using the 8-bit bus and Fast-20-SCSI timing;
- Connection to up to 15 Wide-SCSI devices, or max. seven 8-bit devices;
- Automatic adapter termination via SCSI*Select* utility;
- Modifiable boot device settings;
- Plug & Play support to *SCAM*.

Adaptec supplies the AHA-2940UW with two installation helpers. One of these is EZ-SCSI 4.0 as discussed earlier. The other is called Adaptec *7800 Family Manager Set*. In the installation manual, Adaptec indicates that EZ-SCSI is required for the installation under DOS/Windows 3.1, while the Manager Set is used for all other operating systems (including Windows NT).

Since using a Wide-SCSI adapter only makes sense in conjunction with a powerful operating system, *EZ-SCSI 4.0* is best forgotten at this point.

The installation of the AHA-2940UW should be trouble-free. The statements referring to the PCI slot settings (see Symbios 8250S installation) apply here, too. You should, therefore, be sure to use a PCI Master Bus slot (*PCI Rev. 2.0*).

The ID number for the boot device is '0' by default, while that of the host adapter is '7'. There should be no reason to change these settings, unless, of course, the number allocation by SCAM appears to be useful (see *Plug & Play*).

The termination setting for the adapter (modifiable with the SCSI*Select* utility) is normally set to *Automatic*, i.e., the adapter checks during booting which cables are connected (internal and external, 16- or 8-bit) and changes the termination of the adapter card accordingly. The automatic termination operates satisfactorily, and there should be no reason to switch it off. The termination of the end devices on the respective cables should, of course, be arranged by the user.

3. SCSI in practice

If the termination of the adapter is also carried out manually, the following points should be noted:

With the termination of the Wide-SCSI bus, a distinction is made between *LOW* and *HIGH* bytes. The *LOW* bytes stand for the termination of the original 8-bit wide bus, the *HIGH* bytes, for the newly added lines under Wide-SCSI.

connected devices	High Byte	Low Byte
only external	term.	term.
only internal	term.	term.
external/internal 8 Bit	term.	—
external 16 Bit/internal 16 Bit	—	—
external 8 Bit/internal 16 Bit	term.	—
internal 16Bit/internal 8 Bit	term.	—

Table 3.8.
Termination of the LOW and High byte depending on the connected SCSI cables.

Table 3.8 elucidates in which cases termination is required.

SCSI*Select* allows other settings to be modified in addition to the bus termination. The utility is started with the keyboard combination <CTRL><A>, as soon as the message

```
Press <CTRL><A> for
SCSISelect (TM) Utility
```

appears during the boot-up procedure.

The main settings of the utility are:

- Deactivation of parity checking
- Adapter termination on/off control
- Modifying the boot ID/boot LUN
- On/off switching of Disconnect/Reselect
- On/off switching of *Start Unit* command at the start of a boot-up procedure (see Appendix: *Utilities* DSP220.EXE)
- On/off switching of SCAM support (Plug & Play)
- Modifying Int13 mapping for DOS disks
- On/off switching of BIOS support for bootable CD-ROMs.

The Windows NT driver for the AHA-2940 UW is called AIC78XX.SYS. The installation is straightforward and extensively described in the manual. Unix drivers are not missing either for this adapter.

The complementary version for the Apple Macintosh is called Power Domain 2940 UW. As regards performance and functionality, this is an identically constructed Ultra-Wide SCSI adapter for the PCI slot of the Power PC.

Active termination and software-driven installation is just as self-evident as with the AHA-2940 UW. The external Wide-SCSI socket is not different from that on the adapter for Intel computers; fortunately, identical paths have been followed in this respect.

Also included in both adapter kits are a 68- and a 50-way internal SCSI cable, each having three plugs, so that two Wide-SCSI and two 8-bit SCSI devices may be connected. Regrettably, neither of the two cables complies with the SCAM recommendations, so that they can not be used as part of a Plug & Play setup.

The Macintosh version of the adapter is compatible with the *Remus* software for the MacOS. In conjunction, the software and the Wide-SCSI adapter allow disk arrays to be built (RAID 0, RAID 1, RAID 4 and RAID 5), which allow a considerable increase in data throughput to be achieved (see *Meaningful distribution of data*).

3. 6 Plug & Play

Automatic termination of a SCSI adapter card is not the only function to achieve increasing usage in the light of Plug & Play. The automatic and still correct allocation of ID numbers and termination of the end devices should be feasible features. The following paragraphs discuss how much of this is still wishful thinking, and what has been achieved already.

SCAM

Apart from the quest for ever higher performance, increasing importance is attached to 'ease of use'. The *SCAM* technology (**S**CSI **C**onfigured **A**uto**m**atically) follows in the wake of Microsoft's Plug & Play solutions for Windows 95 to lighten the job of configuring the SCSI bus.

Automatic recognition of connected SCSI devices has been a feature of the SCSI system for many years. Each BIOS and each driver provides a screen message about the devices found on the bus (see Figures 3.25, 3.31 and 3.32), while automatic termination of an adapter card does dot seem to be a serious problem to solve for equipment manufacturers. Because the adapter should use the occupied SCSI connectors to recognize if a byte has to be terminated, and, if so, which byte, even a hardware solution may help out in this case. That software solutions seem to be preferred anyway may be caused by the need for faulty terminations (too many or too few cable terminations) have to be recognized also on end devices to enable an error message to be produced if necessary.

Error-free termination may be recognized with the aid of level measurements if cables are applied having the correct impedance, only active terminators are used, and the supply of the TERMPWR line is forcibly assigned to the host adapter.

Consequently, the following conditions have to be observed to obtain a functional SCAM termination:

- Adapter manufacturers are forced to supply an internal cable with end termination with their SCSI plug-in cards. That allows termination to be omitted on all internal devices.
- The external bus connection of a SCSI adapter featuring SCAM support should be an isolated HD socket; 50-pin with an 8-bit bus; the 68-pin connection is used under Wide-SCSI. Each external connector that complies with this standard should be marked by the SCSI symbol (Figure 3.43).
- To guarantee correct termination of the SCSI devices, the terminator on the host adapter has to be laid out such that it is automatically switched off when an internal and an external connection is made. Wide-SCSI adapters in addition have to make a distinction between LOW and HIGH bytes.
  Correct termination is ensured at the internal side by the wire termination of the inserted cable. As before, the end device at the external side has to be terminated manually. The adapter supplies a warning only if errors occur in this respect.
- The supply voltage on the TERMPWR line is provided by SCSI adapters (only). All terminators are powered from the TERMPWR line (or lines with Wide-SCSI).

Figure 3.43.
The SCSI symbol, which should be prominent on all external connectors complying with the SCAM standard.

3.6 Plug & Play

In addition to termination, duplicate ID number assignment is a frequent source of problems on the SCSI bus. While devices without SCAM functions do not present serious problems with (internal) SCAM termination (the termination being switched off on all internal devices), serious trouble may arise with the allocation of ID numbers.

Any automatic ID number allocation system should take into account that SCSI devices without a SCAM function occupy an ID number that may only be changed manually. As a result, you have to make sure that these numbers are recognized, and that only free ID numbers are assigned to devices having a SCAM function, thus eliminating duplicate assignments.

On the other hand, Plug & Play SCSI devices should also be usable in systems without SCAM, that is, they should have ID settings that may also be used in systems having manual configuration.

To this end, all SCSI devices with SCAM functions are turned into so-called Standard IDs, arranged into device classes allocated by the manufacturer. This allocation is shown in Table 3.9. These Standard IDs may be modified, if necessary, to match the requirements in non-SCAM systems by modifying a jumper setting.

The actual sorting operation with the automated ID assignment runs in so-called *Isolation Mode*. If a SCAM device is in isolation mode, it does not occupy an ID number. In this way, the host adapter is capable of telling if there are non-SCAM compatible devices on the bus. These devices are recognized in isolation mode by their ID number. In that case, the adapter stores the IDs of these devices as 'occupied' and no longer available for allocation.

The ID distribution procedure in a SCAM system follows this pattern:

- The adapter supplies a SCSI Reset, and initiates the *SCAM Select Sequence*.
- All SCAM compatible SCSI devices on the bus go into Isolation Mode.

SCSI ID	Device type
7	Host adapter
6	Magnetic disk drives
5	
4	Streamer or optical devices
3	CD-ROM drives
2	Scanners or printers
1	
0	

Table 3.9.
Manufacturer-assigned standard IDs for different device classes (SCAM devices).

- The adapter reads the remaining IDs (of all non-SCAM compatible devices) and stores them.
- Using the Device identification, the adapter recognizes the device classes, and, as a result, the Standard ID of the first SCAM device. If the Standard ID is not occupied yet, it is assigned to that device. If not, the device gets the next lower ID number, etc.
- All SCAM devices are distributed across the free ID numbers on this way.

In the process of distributing the IDs to SCAM devices, the adapter should take the following requirements into account:

○ If possible, an attempt is made to assign the Standard ID.
○ If this has been assigned already, the device gets the next smaller number which is free.
○ If ID '0' can not be assigned any more, the adapter jumps to the highest free ID.
○ If no free ID can be found, the relevant device is not assigned a number (in practice, this means that more SCSI devices are connected to the bus than the allowable maximum).

An example:
Let's assume that four SCSI devices are connected to the bus: two SCAM devices and two without SCAM functionality.

- Device A is a SCAM compliant CD-ROM drive
- Device B is a had disk without SCAM functions
- Device C is a hard disk with SCAM functions, and
- Device D is a non-SCAM compliant tape streamer

The result of the ID assignment is shown in Figure 3.44. The ID numbers of device B and D are fixed, because devices are involved that do not have SCAM functionality.

Device C is a hard disk drive with Standard ID 6. Because this ID is still available, it is actually assigned.

Device A is a CD-ROM drive with Standard ID 4. Since this number is already occupied by device D, the next lower value is selected for A. However, since number 3 is already in use for device B, A is finally given ID 2.

3. 7 SCSI devices

Isolation sequence	Standard ID	Allocated ID	Set ID	Device
—	—	—	4	D
—	—	—	3	B
1	6	6	—	C
2	4	2	—	A

Figure 3.44.
Result of an allocation sequence with four SCSI devices on the bus, where two are SCAM compliant, and two others have manually set ID numbers. Device A is a CD-ROM drive, B a hard disk drive, C a hard disk drive and D a tape drive.

In particular the assignment of IDs for the hard disks in our example clashes with the practical approach used so far of assigning ID 0 to bootable drives. Because of this, it may be helpful to change the ID of the boot drive, SCAM compliant adapters allowing, of course.

Although such automatic allocation of decreasing ID numbers may help to avoid errors caused by duplicate numbers, it does not enable the priority assignment between individual drives to be affected in any way. However, the priority does not have a particular fixed order in any case, because very few users will worry about it, and instead simply apply consecutive numbering.

Because of this, SCAM technology may be valued as a considerable step towards more ease of use, while also offering SCSI newcomers a high degree of configuration certainty.

3. 7 SCSI devices

After cables, BIOS, software interfaces and SCSI adapters I will endeavour to offer an overview of the various SCSI devices and their potential problems. Before going into detail as regards the individual device classes, some light should be shed on the matter of the specific application areas in which internal and external devices are preferred.

3. SCSI in practice

Internal or external devices?

The circumstance alone that such a question may be asked at all may be considered big plus for the SCSI system. There is no other interface for the connection of peripheral devices that comes anywhere near this performance, while also offering the possibility to connect external devices. With E-IDE, for example, the maximum cable length in PIO Mode 4 is about 30 cm. As a matter of course, external connections are out of the question here.

External SCSI devices are not, in general, modified constructions, but devices for internal use that have been fitted into an enclosure together with two external SCSI connections and an add-on power supply. As a matter of course, external SCSI devices are more expensive than internal ones, simply because an enclosure and a power supply cost money. The additional investment does, however, buy you more flexibility.

External SCSI devices may be used in conjunction with different computers, without complex installation procedures. An external tape streamer, for example, which is to be used for data backup purposes, is typically used on a computer which is due for a backup session. The software installation is then limited to a 'once-only' device driver configuration on each computer. If the computer is started with the tape streamer not connected, the

Figure 3.45. Removable disk drive fitted in an external case. The power supply (top), the actual SCSI device (insertion-type) and the SCSI connecting (left) are clearly recognized.

3.7 SCSI devices

driver reports that the corresponding device was not found, and returns to its normal business.

As regards hardware, the streamer is linked to the external SCSI connector of the computer (where only the last device is terminated as usual), and hooked up to the mains. When the PC is started, the SCSI adapter immediately recognizes the new participant, without having to run any kind of installation, provided the ID number of the external tape streamer does not clash with any of the IDs belonging devices already connected to the bus.

The advantages of external SCSI devices may be summarized as follows:

- ☺ Purpose-oriented device usage on different computers, coupled with minimal installation complexity.
- ☺ Devices are only switched on when actually used (energy saving).
- ☺ If the computer case is too small for extensions, an additional SCSI device may be connected without problems.

The disadvantages are the following:

- ☹ An external device is more expensive than an internal one.
- ☹ Expensive, round SCSI cables are required to connect external devices.

Magnetic disk drives

Computer hard disks are the SCSI devices with the widest use. They are available in different sizes and constructions. For standard storage volumes of up to 1 Gigabyte, most manufacturers supply E-IDE and SCSI versions which are identical with the exception of the interface.

In general, there is no difference in performance between SCSI and E-IDE versions of the same hard disk. If, however, an operating system is used which is capable of accessing multiple drives in parallel fashion, then SCSI has distinct advantages thanks to its

3. SCSI in practice

Disconnect/Reselect functions. In general, several hard disks may be connected under SCSI without problems, while not every drive is guaranteed to function properly on a particular E-IDE controller, or in conjunction with other E-IDE drives.

The disadvantage of SCSI hard disks is that they are, in general, some £25 to £35 more expensive than their E-IDE counterparts.

The mechanical installation of a SCSI hard disk in a computer system is straightforward. The unit is either fitted in a 3.5-inch bay using mounting brackets, or secured in an external enclosure with the aid of spacer bolts.

The electrical connection is also unlikely to cause problems. The internal SCSI cable (pin 1 to pin 1, 68- or 50-way) and the polarized plug (having a bevelled edge) for the supply voltage are simply plugged on to the mating connectors at the rear of the drive unit.

Before finally securing the drive unit, however, it is recommended to have a look at the circuit board attached to the drive unit. As illustrated in Figures 3.48 and 3.48a, this side of the board has two miniature plug connections, and a jumper block is present behind the SCSI connection.

The miniature connections are required when the drive is built into an external enclosure. The connection *J2* is then linked to the LED in the external case, thus enabling any access to the disk to be signalled by a small light. Provided the LED connection of the SCSI adapter is wired to the LED in the computer case (se Figure 3.39),

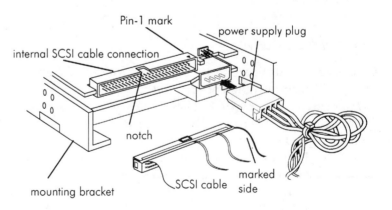

Figure 3.46. Electrical connextions at the back of a SCSI hard disk. These connexions are identical on practically all SCSI devices.

3.7 SCSI devices

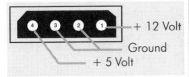

Figure 3.47.
Pinout of the power supply connector.

the LED will light anyway on each disk access, so that the connection to the LED in the external case may be omitted.

The connection marked J5 is connected to ID switch at the rear panel of the external enclosure. The ID number of the device may then be set via this switch (*Remote Control*). This, however, requires jumpers A0-A2 to be removed from the jumper block.

Figure 3.48.
Board side view of a Quantum SCSI hard disk drive. To the right, the main connections, behind them, the jumper block, and, to the left, the miniature plug connections for LED and ID switch.

Figure 3.48a.
The jumper block via which the ID number and termination may be set.

If the hard disk is integrated into the computer case, the jumper block needs closer examination. The jumper marked *TE* (Termination Enabled) switches the bus termination of the drive on and off. A jumper in this location means that the disk is terminated. These settings may, of course, be omitted, or they may not be available at all, if the disk drive is of the SCAM class (see *Plug & Play*).

However, with internal SCSI devices having no Plug & Play assistance we may also make termination much easier. Any computer shop should be able to supply you with a plug-on terminator for 50-way flatcables (see Figure 3.49). If this device is plugged on to the last socket of the internal SCSI cable, none of the internal SCSI devices requires a terminator any longer. The SCAM system exploits the same technique, only the terminator is integrated into the cable.

3. SCSI in practice

Figure 3.49.
Plug-on terminator for 50-way flatcable.

The flatcable terminator shown here is a passive version. Fortunately, it is suitable, according to the SCSI standard as well as my personal experience, for use in a Fast-SCSI system.

A termination implemented in this way does, of course, call for a free plug on the flatcable, which may present a spot of bother with the usual cables having three plugs only. You then have to choose and decide: either invest in a cable with five or more plugs (keep the cable length in mind!), or use tweezers to set the termination jumper on the last device.

Jumpers A0 - A2 have to be set when the drive is installed into a computer case, because they determine the ID number. Quantum, for example, applies the following arrangement for their hard disk drives.

Table 3.10.
Jumper/ID-relations, as used with Quantum hard disks.

ID/Jumper	0	1	2	3	4	5	6	7
A0	0	0	0	0	1	1	1	1
A1	0	0	1	1	0	0	1	1
A2	0	1	0	1	0	1	0	1

0 = jumper not set   1 = jumper set

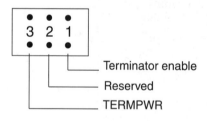

Figure 3.50.
Jumper block on a Maxtor disk. Be sure not to mistake the Termination for the TERMPWR supply, or vice versa!

Care should be taken when a jumper is used to actuate the TERMPWR line on a hard disk drive. On some Maxtor disks, the jumpers for *Termination Enable* and *TERMPWR* Enable are very close together (Figure 3.50). The TERMPWR line is normally powered by the host adapter. Only in exceptional cases, if the adapter is unable to do so, a SCSI device should be allowed to take over the power supply of this line. Under no circumstances must the TERMPWR line be powered by several sources, because that may lead to unexpected errors.

218

If the purpose of building a hard disk drive into an external enclosure is to allow the unit to be used with different computer systems, acting as a transport medium for large amounts of data, then problems may brew under the DOS and Windows operating systems.

As explained in the section *SCSI BIOS boot helper*, the addressing of logic blocks has to be converted to physical cylinders, heads and sectors of the hard disk if the SCSI disk is to work in conjunction with the PC/AT BIOS (applied under DOS and Windows). Depending on the use of different mapping algorithms, it may happen, under unfavourable conditions, that a magnetic disk may not be legible when attached to another SCSI adapter, simply because the way in which a SCSI adapter mirrors logic blocks on to cylinders, heads and sectors, was not fixed as yet.

A magnetic disk may be used on a SCSI adapter that uses a different mapping after re-formatting only, which does not help to promote the use of an external hard disk drive as a data transport medium.

But here, too, the new SCAM technology may offer advantages. SCAM compliant host adapters are subject to the obligatory use of an algorithm which recognizes foreign mapping and puts an adapter in a position to copy it. The latest adapters from NCR/Symbios prove that it works.

Large hard disks
From a capacity of about 3 Gigabytes onwards, E-IDE hard disk drives are a bit thin on the ground. Above this limit, SCSI drives are practically alone. Today, the largest among them achieve storage capacities of up to 9 Gigabyte (*Barracuda 9* from Seagate: 9.1 Gigabytes formatted) at a physical size of 3.5 inch.

These disks spin at 7,200 rpm, have average access times of 8 ms, and feature either a Fast-20-SCSI interface for the 8-bit bus or the 16-bit Wide-SCSI bus, or a Fibre Channel interface (see *SCSI-3 — an outlook*). Meanwhile, SCAM functions are implemented by default on these drives.

Figure 3.51. (next page)
Datasheet published by Seagate, providing a performance overview of the Barracuda series hard disks.

3. SCSI in practice

Barracuda 3.5 INCH

Fast, Fast Wide and Single Connector Ultra SCSI and Fibre Channel Interfaces

		ST15150N ST15150ND ST15150W ST15150WD ST15150WC ST15150DC Barracuda 4	ST15150FC Barracuda 4	ST19171N ST19171ND ST19171W ST19171WD ST19171WC ST19171DC Barracuda 9	ST19171FC Barracuda 9
Capacity					
	Unformatted Mbytes	5,062	5,062	11,700	11,700
	Formatted Mbytes (512 bytes/sector)*	4,294	4,294	9,100	9,100
Interface					
		Fast and Fast Wide SCSI-2	Fibre Channel (dual port)	Fast and Fast Wide Ultra SCSI	Fibre Channel (dual port)
Performance					
	Internal Transfer Rate, ZBR (Mbits/sec)	47.5 to 72	47.5 to 72	75 to 120	75 to 120
	External Transfer Rate (Mbytes/sec)	10/20	100	10/20/40	100
	Multisegmented Cache Buffer (Kbytes)	1,024	1,024	512/2,048	512/2,048
	Track-to-Track Seek, Read/Write (msec)	0.6/0.9	0.6/0.9	0.6/1.1	0.6/1.1
	Average Seek, Read/Write (msec)	8/9	8/9	8/9.5	8/9.5
	Maximum Seek, Read/Write (msec)	17/19	17/19	19/20	19/20
	Spindle Speed (RPM)	7,200	7,200	7,200	7,200
	Average Latency (msec)	4.17	4.17	4.17	4.17
Configuration/Organization					
	Discs/Data Surfaces	11/21	11/21	10/20	10/20
	Servo Heads	1	1	Embedded	Embedded
	Bytes per Track	49,350 to 74,900	49,350 to 74,900	70,830 to 111,460	70,830 to 111,460
	Sectors per Drive (512 bytes/sector)	8,388,315	8,388,315	16,992,187	16,992,187
	Cylinders	3,711	3,711	5,333	5,333
	Recording Method	RLL (1,7)	RLL (1,7)	PRML (0,4,4)	PRML (0,4,4)
Reliability/Data Integrity					
	MTBF (power on hours)	800,000	800,000	1,000,000	1,000,000
	MTTR (hrs)	depot	depot	depot	depot
	Service Life (yrs)	5	5	5	5
	Recoverable Read Errors per Bits Read	10 per 10^{11}	10 per 10^{11}	10 per 10^{11}	10 per 10^{11}
	Nonrecoverable Read Errors per Bits Read	10 per 10^{14}	10 per 10^{14}	10 per 10^{14}	10 per 10^{14}
	Seek Errors (per seek)	10 per 10^8	10 per 10^8	10 per 10^8	10 per 10^8
	Limited Warranty (yrs)	5	5	5	5
Power Requirements					
	+12 VDC ±5% (amps typ)	0.75	0.98	0.7	0.88
	(amps max)	2.2	2.2	2.2	2.2
	+5 VDC ±5% (amps)	0.65	1.33	0.8	1.16
	Power (Idle watts)	12.5	16.3	12.4	16.3
Environmental					
	Operating Temperature (°C)	5 to 50	5 to 50	5 to 50	5 to 50
	Nonoperating Temperature (°C)	–40 to 70	–40 to 70	–40 to 70	–40 to 70
Physical					
	Height (inches/mm)	1.63/41.4	1.63/41.4	1.62/41.4	1.62/41.4
	Width (inches/mm)	4/101.6	4/101.6	4/101.6	4/101.6
	Depth (inches/mm)	5.75/151.6	5.75/151.6	5.75/146.1	5.75/146.1
	Weight (lb/kg)	2.3/1.04	2.3/1.04	2.3/1.04	2.3/1.04

* Capacity calculated with typical sparing. Sparing is user optional.

Corporate Headquarters
Seagate Technology 920 Disc Drive, Scotts Valley, California 95066, 408/438-6550

Asia/Pacific
Seagate Technology International 202 Kallang Bahru, Singapore 339339, 65/292-6266

Europe, Africa, Middle East
Seagate Technology 62 bis, avenue André Morizet, 92643 Boulogne-Billancourt Cedex France (+33 1) 41 86 10 00

Seagate, Seagate Technology and the Seagate logo are registered trademarks and Barracuda and the Barracuda logo are trademarks of Seagate Technology, Inc. ©1995 Seagate Technology, Inc. All rights reserved. Specifications and product offerings subject to change without notice. Printed in USA.

Publication number 1409-007   Printed on recycled paper

Seagate
The Data Technology Company

3.7 SCSI devices

Not just because of their high price, these hard disk drives are not really suitable for use in desktop systems. The high disk rotation speed in particular makes fitting such a drive unit into an enclosure which is close to the user rather an unwise recommendation. Obviously, some extra noise can not be avoided at such high disk rotation speeds — in plain language, these hard disks produce a racket.

Because of this, the main application area is not only limited to server systems running in separate, air-conditioned rooms, and annoying the occasional service engineer only with their noise production.
   Considering the application area and the requested high data security on the one hand, and the larger allowable cable lengths on the other, it is not surprising to note that these disks are often offered for differential SCSI (see Appendix).

Disks with a capacity of between 1 and 4 Gigabyte may, however, be of interest to the average user. Meanwhile, the storage cost per megabyte has dropped to a record low of about 12p (July 1996) for all hard disk sizes.

Removable disk drives

With SCSI removable disk drives whose media volumes may lie between 44 Mbyte and 2.6 GByte, the data throughput is rarely as high as with hard disks, so that Wide-SCSI or fast-20 timing are not significant.

Syquest's drive units made the company a long-time market leader in magnetic removable media. Lately, however, Iomega has become a formidable opponent thanks to the success of its drives. Iomega's Zip media look very much like ordinary 3.5-inch disks, while the drives achieve a data transfer rate of about 500 kByte/s, peak values approaching 1 Mbyte/s. A 100-MB medium costs about £13. The medium cost for an EZ-135 drive from Syquest is about the same (135 Mbyte), although the data transfer rate is about two times as high at 1.1 MByte/s. Maximum (burst) transfer rates for the EZ-135's read and write operations are of the order of 2.2 Mbyte/s.

3. SCSI in practice

Syquest packs its data carriers in transparent (3.5-inch) cartridges which enable access to the data carriers via a slide mechanism at the front side as soon as the cartridge is inserted into a drive unit.

Figure 3.52.
Typical form of Syquest removable media, here, an older 5.25-inch version is shown.

The older 5.25-inch drive units whose memory volumes of 44 and 88 Mbyte per medium no longer meet today's technology requirements are still supplied by Syquest because they are in very wide use.

The Jaz drives from Iomega use a similar cartridge shape. A cartridge contains two magnetic disks which give this medium a memory volume of 1 Gigabyte. As regards data transfer rates, Jaz drives are currently way ahead of the competition. The data transfer measured as an average of read and write access operations achieves a solid 1.6 MByte/s, while maximum values with read access operations may exceed 6 Mbyte/s. Writing to such a disk clearly takes a bit longer — here, even the peak values remain below 1.5 MByte/s. The storage costs are about 8p/MByte.

Removable disk drives are available as internal and external units, where the SCSI interface is dominant throughout. Low-cost solutions are available for the printer port (low data throughput) as well as for E-IDE interfaces (only internal devices). However, the E-IDE interface also requires a driver to be able to handle a removable disk drive, and the situation is not at all clear in this respect.

External SCSI drives in particular are frequently used for data backup and data exchange purposes, because they are simple to install on different computer systems, if and when required. The device drivers supplied by the manufacturer are implemented on the ASPI interface.

At least the Syquest drives and the Iomega Jaz drive may be addressed using drivers for hard disk drives, where the SCSI adapters from NCR/Symbios Logic detect whether or not a medium is installed in the drive when the system is started.

If a medium is installed, the removable disk drive unit is treated like a hard disk, so that its medium may not be changed without having to restart the computer (otherwise, the newly inserted medium would be at risk of having data overwritten, the computer still assuming that the original medium is in the drive).

If, on the other hand, the removable disk drive unit is not mounted (i.e., no disk is inserted) during the computer's start-up sequence, the Symbios adapters create a dummy drive unit after a delay of about 30 seconds (multiple drive letters are possible for different partitions, see *Symbios 8250S*). The dummy device affords medium-changer operation without problems.

In this mode of operation, a check is run with each medium access, to see whether the same data carrier is still in the drive. In this way, a change of medium is recognized in all cases. Although the continuous checking does reduce the data throughput a little as compared with operation as a 'hard disk', do remember that you are looking at a removable disk drive, so that you have to assume that a medium change is aimed for during normal operation.

Magnetic removable disks have competitors in removable media on magneto-optical basis (see *MO drives*).

CD-ROM drives

CD-ROM drives have been the protagonists of a market boom over the past few years. These days, hardly any computer is shipped without such a disk player.

No wonder looking at the torrent and massive volume of today's software packages. Which user is still prepared to feed his/her computer with 20 or 30 floppy disks, if the same software installation is carried out practically by mouse click via CD-ROM?

Because of the aggressive pricing policies of computer manufacturers, CD-ROM drives having an IDE/ATAPI or an E-IDE/ATAPI interface are the most frequently seen. SCSI drives are considered a luxury here.

3. SCSI in practice

Figure 3.53.
Six-speed CD-ROM drive from Plextor.

As a rule, they are also more expensive, which may, however, be explained by better-quality hardware. A glass lens (instead of plastic) for laser beam focussing, an effective dust protection, or a mechanism for automatic lens cleaning, all these features are available as a matter of course on SCSI CD-ROM drives, just as they are invariably missing from low-cost drives.

Cutting through all interface restrictions, two basic drive constructions may be distinguished:

- the CD-ROM arrives in the drive via a slide tray, or

- the user is forced to employ a so-called *caddy*, which is a kind of cartridge for CD-ROMs.

The CD tray is without doubt the more user-friendly solution. The caddy alternative is more expensive but mechanically more robust. Most CD-ROM drives employ a tray mechanism.

3.7 SCSI devices

As regards the installation of CD-ROM drives, there is nothing new to note: device class drivers, whether CAM or ASPI, are supplied with each adapter, while termination, cable connection and ID number setting follow the familiar rules. I also mentioned already that DOS requires a system extension called MSCDEX.EXE (see *CD-ROM.SYS* in section 3.5).

Figure 3.54.
A CD-ROM caddy.

As far as the performance of the devices is concerned (not mentioning the different interfaces), large differences exist. Originally, CD-ROM drives used a disk rotation speed which was derived from audio CD players, resulting in a data speed of 150 kByte/s. Today, some manufacturers supply drives achieving ten times the original speed, resulting in data transfer rates of 1.5 Mbyte/s. The current standard for CD-ROM drives seems to be 4 or 6 times the original disk speed, high expectations being fulfilled by drives with eightfold speed and more.

The data transfer rate is, however, not the only performance criterion which should be kept in mind. The access time, that is, the time that elapses until the laser is positioned above the requested data, has an equally large effect on the total performance of the drive. Also, in addition to the data throughput, the more or less sophisticated error correction methods should be taken into account. If a scratched CD-ROM has to be read, it is the quality of the internal error correction that decides between success and failure.

With SCSI drives, another purchase criterium to consider is the SCAM capability of the devices. *Plextor* is one of the first manufacturers to bring SCAM compliant CD-ROM drives on the market, thereby further simplifying the installation.

One conclusion drawn from the previous discussion could be that the majority of computers to be sold with a CD-ROM player use the (E-)IDE/ATAPI interface to address the drive (this is likely to be true), where, apart from a little dust, no performance degradation is to be expected as compared with the more expensive SCSI drives (this is equally likely to be false).

One of the main application areas of PCs having a CD-ROM drive is Multimedia, where photos, videos, texts, graphics and voice reproduction are integrated to such an extent that high-impact, interactively controlled programs are created which draw heavily on computer power.

During individual processes, such software often requires access to files which are located on different data carriers (CD-ROM, hard disk). It is precisely in this respect that the SCSI system is able to bring out all of its performance. Here, too, the *Disconnect/Reselect* functions allow data from a CD-ROM and a hard disk to be conveyed practically in parallel to the processor.

This is much more difficult to implement on a computer with an E-IDE interface. If the hard disk and the CD-ROM drive are connected as Master and Slave to a single E-IDE channel, they slow each other down. The data from one medium may not be transmitted until a command for the other has been completed. Even if the CD-ROM drive and hard disk drive are switched on to two separate E-IDE channels (if available in the PC), this combination is still noticeably slower than the SCSI competition. A further complicating factor in the 2-channel E-IDE arrangement is that it is no longer possible to integrate a second E-IDE disk into the system (yes, as a slave device, but that would be the worst of all conceivable solutions).

The Macintosh computer uses SCSI CD-ROM drives only, which are integrated via the *Apple CD-ROM* driver. Unfortunately, this driver works with 'proprietary' CD-ROM drives only (supplied by Panasonic and Sony). For Apple users, that means financial and technical disadvantages. The SCSI CD-ROM drives for the PC market are, in general, cheaper, and technically usually one step ahead of the current Apple drives (Apple is slow to follow the general trend towards higher disk speeds). Unfortunately, none of the drive manufacturers offers a CD-ROM driver for Apple (rumour has it, though, that you might be successful by applying to the manufacturers). Assistance may be provided by the *FWB CD-*

ROM Tool, whose price does however, negate the saving obtained by purchasing a CD-ROM drive for the PC market.

The interaction of drive unit and CD-ROM driver is the decisive factor when it comes to application areas of the unit. One question that was often heard some time ago: "*Does the driver support Multi-Session Photo CDs?*" is no longer relevant — this hurdle is easily jumped over by all of today's drivers. The criterium is now the reading of Video CDs and CD-I disks. To be able to read Video CDs, the drivers have to comply with the CD-ROM-XA/Form 2 standard. Reading CD-I disks requires additional drivers in any case because these disks, like audio CDs, do not have a file system.

CD changers

CD changers are CD-ROM drive units that have a storage container for several disks, which may be accessed individually by selecting them. They can not be used in conjunction with normal CD-ROM drivers because the medium change operations require them to handle SCSI commands from the device class of *Medium Changer devices*.

The first CD changers (Pioneer, NEC) were able to hold up to 7 CDs, and they were available as external drives only with a SCSI interface, probably because such a combination of a drive and a disk container could not be fitted in any disk drive bay. Recently, CD changers have become smaller and are limited to three CD-ROMs, which allows them to be fitted in a drive bay. Common to all devices is the fact that only one scanning mechanism is available, so that it is not possible to read several CD-ROMs at the same time.

A CD changer is controlled via a SCSI ID and a corresponding number of LUNs. Each CD slot is assigned a drive letter by the DOS, Windows (incl. '95) and OS/2 operating systems, so that you have to know which CD-ROM is in a particular slot to be able to select the disk under the right letter.

A more elegant solution is found on the Macintosh computer (here, manufacturers are less tight-fisted than with individual drive units). After a CD-ROM drive has been mounted, each disk may be addressed via its volume label, clearly lightening the load on the user's short-term memory.

CD writers

The units from the class called *Write-Once Devices* are exclusively of the SCSI type — CD writers controlled via another interface (E-IDE or similar) have not emerged yet. None the less, many of these devices are unable to properly employ one of the main advantages of the SCSI system, namely the use of several different devices on a single adapter.

This is because the data flow may not be interrupted during the actual write operation (the 'burning' of the disk). Should that happen, the CD-R may be thrown away because it is unusable. Unless the cache memory in the drive has a size that allows interruptions in the data flow to be buffered without problems, most manufacturers resort to the use of a separate SCSI adapter (usually for the ISA bus), which is then exclusively used to drive the CD writer — big sales for *narrow-track adapters* (see the theory chapter). Such a SCSI adapter costs less than a proper memory outfit, the more so because low-spec versions are involved mostly which lack arbitration features, and are capable of driving a single SCSI device only. The cost saving achieved by cutting a decent cache memory is, however, negated by a troublesome installation procedure. A free memory address, a free IRQ line and a free DMA channel, all of this is known from the installation of a full-featured SCSI adapter. Furthermore, you have to install an extra CAM or ASPI driver on top of the device driver (different bus system, different manufacturer), which may also cause disappointment.

SCSI implemented in this way means that all advantages of the system are thrown over board.

If I may be allowed to give you a tip: should you consider buying a CD writer, and come across an attractively priced solution that does, however, allow the writer to be used on a separate adapter only, don't go for it!

Although the majority of newly arrived devices on the market is now capable of working together with other devices on a single SCSI adapter, many are still unable to do so.

The CDD2000 from Philips as pictured here works happily in my PC system, together with three hard disks and a CD-ROM drive attached to an AHA-2940. It writes all known CD formats (including CD-DA, CD-I and Video CD) at double speed. Its cache memory has a size of 1 MByte.

Figure 3.55.

CD writer type CDD2000 from Philips (here, the external device version is shown). The unit is also marketed by HP and Plasmon running their own firmware, and has produced a price shock in the relevant market section.

Drivers are available for DOS, Windows, MacOS and Unix. At a street price of about £475 for an internal device without software, this or similar equipment leads the way towards a breakthrough on the mass market.

When used as a regular CD-ROM drive, the CDD2000 achieves quadruple speed operation, albeit that its average access time is a mediocre 400 ms or so (as compared with 120 to 150 ms for most quad-speed drives).

PD drives

PD stands for *Phasewriter Dual* and refers to a drive type which is capable of reading and writing a cartridge-held 650-Mbyte medium using *Phase-Change Technology*, as well as acting like a regular CD-ROM drive.

Phase-Change Technology refers to a pure optical technique whereby a laser heats a recording layer at different intensity levels. Depending on the degree of heating, the laser-covered areas of the medium remain either in a crystalline or an amorphous state after cooling down. These different states cause different angles of reflection to a light beam emitted by a scanning laser (operated at considerably less power than with writing). In this way, binary information may be stored on to such a medium. In contrast with the write operation on a CD writer, the information on a PD medium may be overwritten up to 500,000 times.

3. SCSI in practice

Figure 3.56.
Internal and external version of the PD drive from Panasonic. Re-recordable media on the top of the devices, and a CD-ROM inserted in each drive.

The fact that CD-ROMs may also be read by such a drive brings us nearer to the ideal situation in which CD-ROMs may be overwritten as many times as one would like.

Although this twin-standard drive is still different from the ideal because it uses two different media, it does appear to be a highly interesting SCSI device.

As with nearly all new introductions on the market, the PD drive is initially shipped with a SCSI-2 interface, and it may serve as an example from the class of *Optical Memory Devices*. After all, this class is capable of handling different media because the media type is checked with each access using the *Mode Sense* command.

The performance of this new device is not impressive, neither in the CD-ROM range, nor as a Phase Change Medium.

CD-ROMs are read at 600 kByte/s (normal quad-speed rate) and an average access time of 200 ms, while the data transfer rate for the Phase Change medium indicates a reading speed which is two times higher than the writing speed (write rate approx. 350 kByte/s, read rate approx. 750 kByte/s).

The list price of an internal drive and one medium included is currently around £375. The external version is about £85 dearer. The device comes with drivers from the Corel driver kit which allow application under DOS and Windows, but also with a Mac computer. As with all optical storage media, the driver *OPTICAL.SYS* is applied under OS/2.

The drive is managed under one ID and two LUNs, so that two drive letters are assigned to it, the first representing the PD medium, the second, the CD-ROM drive.

It remains to be seen whether this dual medium will gather sufficient buyer's interest. If the street price drops more than £100 under the list price, the device could meet with some interest from private PC users.

MO drives

As indicated by their name, magneto-optical (MO) storage devices employ a combined magnetic/optical technique. The actual data carrier is a magnetic material. In contrast with traditional magnetic storage devices, however, the data carrier is locally heated to a temperature at which the Curie level of the magnetic layer is exceeded. In this state, a small magnetic field is sufficient to make the magnetic elements in this areas point into a desired direction. If the area cools down again, the polarisation is retained, and can not be changed except by applying very strong magnetic fields. Information may be stored by making individual 'memory cells' point in different directions, depending on their logic value (0 or 1). This information may be read again with the aid of a polarized laser beam, which is reflected in slightly different directions by the aligned magnetic areas.

A characteristic feature of MO media is that their data throughput for write operations is 2 to 3 times smaller than for the read operation. This is caused by the fact that an MO medium is written to in two passes (three during a test run). To enable a high recording density to be achieved, the entire data carrier is exposed to a constant magnetic field, while a laser beam is used to heat exactly those points that need to be affected. A laser is used because it can be focussed with far greater precision than a magnetic field. In this way, all zeroes are written on to a track during the first spin. Next, the magnetic field is reversed, and the desired areas are turned into the other direction, i.e., written to with ones. A third pass is often appended to verify the previously written data. Later, a single rotation is sufficient for the read operation.

Although MO media suffer from a relatively low data transfer rate, they do offer better data security than normal removable media, firstly, because a magnetic field of average strength can not harm the data, and secondly, because a head crash causing damage to the drive and the disk is impossible thanks to the use of a laser beam.

The data transfer rates of the currently available 3.5-inch MO drives are 400 kByte/s for write operations, or 1.2 MByte/s for read access. The 5.25-inch drives are a little faster, offering peak values of about 700 kByte/s or 2.1 MByte/s for the write and read operation respectively. Media volumes of 128 Mbyte up to 640 Mbyte are available for the 3.5-inch version, while the 5.25-inch version may use media volumes of up to 2.6 GByte.

In general, it may be said that MO drives are more expensive than magnetic removable disk drives (Syquest, Iomega) when it comes to purchasing. The storage media (assuming equal sizes) are, however, cheaper. Comparing, for example, drives with a media size of 230 Mbyte (Syquest on the magnetic side, Olympus of the MO side), costs shift in favour of MO drives as soon as more than ten storage media are required. At a smaller number, magnetic removable media are more cost effective.

Gigabyte MO drives costing more than £850, these giants play in a different league, and may not compared price-wise with magnetic removable drives, although they come very close as regards performance.

3.7 SCSI devices

All MO drives currently on the market use a SCSI interface. Because of this circumstance, most drives may be supplied as internal and external devices. In the SCSI Type-Byte Identification system, MO drives report with the ID *07* by default — the ident of an optical system. On many devices, this identification may be changed to *00* by means of a DIP switch or a jumper block, so that they identify themselves as a magnetically recordable disk, and may be integrated into the system using the traditional magnetic disk drivers. Most manufacturers supply MO drivers tailored to their drive units, so that installation is not generally a problem. Under Windows 95, an MO drive is integrated as a regular removable disk drive. Should the formatting cause problems, it is recommended to make use of the low-level formatting utilities supplied by the adapter manufacturer.

For use with Macintosh computers, certain devices (for example, the M2512A from Fujitsu) have to be set to 'Mac' mode with the aid of a jumper.

Furthermore, the write cache and the verification of the finished write operation may be switched on and off on this equipment.

The fact that the M2512A drive is capable of powering the TERMPWR line should be hailed with some reservations. It is a jumpered setting, and should be left 'off' for all the familiar reasons. The device is shipped with termination switched on. This setting should be retained only if the device is the last on the bus.

Key #	Function	Key On	Key Off
01	Write cache mode	Enabled	Disabled
02	Not used		
03	MacMode	Enabled	Disabled
04	Write verify	Disabled	Enabled
05	SCSI type	SCSI 2	SCSI 1
06	SDP (for next)	Enabled	Disabled
07	Not used		
08	Factory test (not to be set by user)	Enabled	Disabled

Table 3.11.
Operating modes besides the usual SCSI settings, offered by Fujitsu's M2512A.

Tape drives

The use of tape streamers is certainly the cheapest way of backing up data. Because of the different recording methods (see theory chapter), the different storage capacities of tapes and the resulting lack of compatibility, you should give some thought to the requirements you want the drive to fulfill before buying one.

3. SCSI in practice

Tape drives are eminently suited to securing complete data carriers by allowing an automatic backup to be made when your work is done. The strength of these tape-based media lies in the storage of large data streams (hence tape **streamer**). They are, however, less suitable for the storage and transport of smaller, selected chunks of data. New data may only be appended at the end of available data, and rewinding right up to this tape location increases the access times to the minutes range, where other data carriers score milliseconds. In case individual data chunks are to be read later, quite a bit of tape winding and rewinding is in order. On the other hand, if data with a volume of the order of several hundred megabytes is stored, and read in one go at a later time, those annoying tape rewinding operations are virtually absent.

The transfer rate at which a tape drive is able to *stream* if it is not interrupted by tape winding and rewinding, depends on the recording method used, the interface and the drive type. At the low end of the scale we find drives that use the floppy disk connector inside the PC for the data exchange, and actually match the data throughput of about 3 MByte/min. achieved by floppy disk drives. At the top of the scale you may come across DAT streamers with a SCSI interface which achieve a data throughput of up to 40 Mbyte/min, which may be a competitive value already to many CD-ROM drives.

As a matter of course, these differences in performance are invariably linked to obvious price differences.

A general disadvantage of all tape drives (irrespective of the price category) is that they do not record touch-free, the applied recording methods being the same as those known from audio and video recorders. Having been used for some time, the tapes are worn (if you haven't witnessed tape spaghetti beforehand), and the magnetic heads are stained as a result of the tape friction.

At the low end of the price and performance scales we find the familiar QIC-80 streamers which employ a 0.25-inch wide tape with a storage volume of 120 Mbyte (*QIC* = *Quarter Inch Tape*). These drives use the serpentine recording method (see *Sequential Access Devices*) which 'guarantees' a low transfer rate.

3.7 SCSI devices

The QIC standard was (and is) frequently updated and extended to boost the performance of the drives in respect of data transfer rates and storage volumes. QIC3010, 3020, a new 'standard' is announced nearly every month. The first devices with a SCSI interface are available from Tandberg, where, however, manufacturer-specific solutions are involved which are not compatible with any other drive.

The *helical-scan recording method*, known from video recorders and DAT cassettes (*Digital Audio Tape*), was copied for DAT tape streamers, and a standard devised for these drives. Helical-scan recording in conjunction with an additional error correction and the *READ after WRITE* principle (possible through the use of separate read and write heads) have resulted in DDS-1, the standard for *Digital Data Storage*. DDS-2 is downward compatible and operates with a higher data density which allows a higher data transfer rate. DDS-3 has been announced.

Figure 3.57.
Actual size of a 90-metre DDS DAT tape.

DAT tapes are available in the following lengths: 60, 90 and 120 m, with equally rising storage volumes: 1.3, 2.0 and 4.0 GByte. Because all tape streamers employ additional compression methods (compression factors between 1.5 and 2), the actually available volume for data storage is increased by these factors. Considering that a DDS compliant DAT tape costs about £6, the storage cost of 0.03p per MByte is low beyond competition.

The cost of a corresponding drive unit having a SCSI interface is of the order of £400 and up. Some drives may already feature SCAM support, and achieve data transfer rates of up to 800 kByte/s. The software that comes with these drives comprises backup programs in addition to SCSI drivers for all current operating systems.

Tape suppliers indicate a lifetime of 2,000 and more head contacts or 100 complete backups for their products. That the magnetic heads may also require some care after wearing out several tapes is usually hushed up.

With QIC tapes, the error correction mechanisms allow damaged areas with a length of up to 7 mm to be recovered. Because of the higher speed, that figure is reduced to 3 mm with DDS DAT tapes. Which of these recording methods is the safer of the two is not certain, but a nice tape tangle is sure to exceed both values.

Scanners

Any PC owner working with graphics programs from time to time will reach a point at which he/she will wish to be able to scan relatively small images into the program (and larger images, of course, as requirements grow), Next, these scans may be edited and modified in any way you like using the program. The available range of scanners includes handheld scanners costing less than £40 right up to high-end drum scanners for phototypesetting bureaus, costing over £40,000.

I prefer not to discuss the different applications and performance of various scanner constructions, limiting myself to look at the SCSI attachment of affordable (flat-bed) scanners, and point out any problems that may arise with TWAIN drivers.

While hand scanners are usually supplied with proprietary insertion cards, flat-bed scanners (price range £250 to £800) employ the SCSI bus to copy scanned data to the PC. The command set available for the device class called *Scanner Devices* ranges from starting the scan operation right up to defining the scan window, i.e, the picture window which has to be scanned in the end.

PC operating systems are not geared to processing and forwarding scanner data to the various picture processing programs, so that the scanning of an image and the integration of the scanned results into an application are separate steps in a process.

3.7 SCSI devices

The scanner software enables the picture section and other preset values to be defined and the scan operation to be started. Once the image data are available, they are saved into a file. Next, the image processing program is opened with which the necessary correction and enhancement work may be carried out. Once the results are in accordance with the user's requirements, the image is saved once again. Next, a DTP (desk-top publishing) program is called which is going to be used to fit the processed image into a publication. If corrections appear to be necessary, the whole process has to be repeated.

Twain

To enable these production steps to be simplified as well as integrated, an interface standard was developed by a number of scanner manufacturers under the name Twain (*Toolkit Without An Interesting Name*). Twain was designed to ensure a better link with the user programs (initially, image processing programs, but also DTP, archiving or DTP applications), so that all individual processing phases could be controlled from one of the programs via *TWAIN* modules.

The SCSI device driver and the *TWAIN* module are generally supplied by the scanner manufacturer, who often includes a (narrow-track) SCSI adapter in the kit. A scanner having a SCSI interface may be driven via the host adapter fitted in the computer. If a free IRQ line, DMA channel and a corresponding base address are available, it may be worthwhile in some cases to fit the narrow-track adapter (i.e, adapter without arbitration) as an add-on to the host adapter, and use it to control the scanner, because some scanner manufacturers have optimized the *TWAIN* module to the card that comes with the scanner (not a satisfactory solution, though, looking at the overall system configuration). In some cases, too, the use of the narrow-track adapter results in increased data transfer performance.

Suitable and flawlessly operating drivers are no longer a problem with scanners, so that a manufacturer may get bonus points by offering a decent level of support service. Here is the WWW site of Hewlett-Packard, the manufacturer of the Scanjet series:

```
http://www.hp.com/cposupport/indexes/swin_dos.html
```

3. SCSI in practice

Digital cameras

I do not intend to define a new device class here, I merely wish to point out that digital photography is on the rise and apparently well aware of the capacities of the SCSI bus, integrating the bus into its total concept.

The difference between 'regular' photography and the use of digital cameras is that the former relies on transferring photographic images on to a light-sensitive film which is chemically processed to make the picture visible. With digital photography, the camera contains a sensor instead of a light-sensitive film. This sensor may be compared to a matrix consisting of individual pixels, whose number per line and matrix column defines the achievable resolution. The Kodak DSC system, for example, operates with a sensor which is capable of resolving $3,060 \cdot 2,036 \approx 6 \cdot 10^6$ pixels at a colour depth of 36 bits. The digital image, that is, the brightness and colour information represented by $6 \cdot 10^6$ pixels, has to be stored on a data carrier (at the highest resolution, this corresponds to about 18 MByte), so that the image may be retrieved and processed later. The data carrier is either a memory card in format or a small hard disk.

And what does SCSI have to do with this? Simple: some cameras that do not have an internal data memory copy their image data directly to the PC running a picture processing program, via a SCSI cable. With others, the image memory is transferred in the same way later. As matter of course, cameras without an internal memory are only suitable for use in studios because of their limited cable length. This need not be a disadvantage, however, because a computer monitor may be used at the same time for image viewing. With some cameras, the entire image capturing technology may even be controlled via a special TWAIN module and the SCSI link.

From a technical point of view, SCSI only helps to forward data from one data carrier to another. Thanks to the external connection and the high data transfer rate, SCSI does, however, offer possibilities to a new technology which are difficult to realise otherwise.

Digital camera model PDC-2000 from Polaroid, for instance, supplies 1600 · 1200 pixels at a colour depth of 24 bits.

It features a /SCSI connection (a card is inserted into the camera, and the other side of the cable has a SCSI plug). Drivers for the Macintosh and Windows are available, the latter using the ASPI interface.

Figure 3.58.
Polaroid digital camera type PCD-200 with SCSI-2 connection.

3. 8 External SCSI adapters

The SCSI systems examined and described so far were all built according to the same structure: inside the PC sits a SCSI adapter to which SCSI devices of any kind may be connected, internally or externally.

How then does one integrate SCSI into a computer that has no room for a traditional SCSI adapter, for example, a laptop or a notebook PC? Should such systems forfeit SCSI functionality if they do not have a SCSI adapter integrated on the motherboard? Two solutions with different levels of performance are available to solve this problem.

SCSI and PCMCIA

The easiest accessible and at the same time most powerful extension slot for notebook PCs and other portable computers is the slot.

A large variety of insertion cards is available for this external extension slot — ISDN adapters, modems, memory extensions, sound cards, and ... SCSI adapters.

Adaptec calls its solution *PA-1460 SlimSCSI*.

3. SCSI in practice

*Figure 3.59.
PCMCIA SCSI
host adapter type
APA-1640 from
Adaptec.*

The adapter being outside the computer case already, it is, of course, only possible to connect external SCSI devices. According to the termination rules, a /SCSI adapter forms the end of the SCSI bus. Consequently, a permanently fitted, active terminator is applied to terminate the bus. As usual, ID number 7 is reserved for the adapter, leaving numbers 0 through 6 available for devices.

ID number 0 does not have special significance in this SCSI system, because it is not possible to boot from connected SCSI devices. It is, therefore, not necessary to allocate a boot device ID. Furthermore, it should be noted that the adapter does not supply the TERMPWR line voltage, although it does power its own terminator. According to Adaptec, this arrangement prevents undue loading of the battery in the notebook. A bit far-fetched, this explanation, because external devices when connected to the notebook will be run off the mains, and no user will then be so stupid as to run his/her notebook off the battery instead of off the mains power supply.

The need for external powering of the TERMPWR line clearly restricts the application possibilities of the *APA-1640 SlimSCSI*:

☹ Only external SCSI devices may be connected, and, because of that, the adapter may only be used if at least one of the devices is capable of providing the supply voltage on the TERMPWR line.

3. 8 External SCSI adapters

The Fujitsu MO drive type M2512A is capable of supplying the TERMPWR voltage, as are some Maxtor hard disks. On CD-ROM drives, tape streamers and removable disk drives, this option is rarely found.

The main features of the APA-1460 SlimSCSI are as follows:

- synchronous data transfer and Fast-SCSI timing
  (allows a data rate which may only be used between external devices, the bus permitting only 2 Mbyte/s)
- PIO Mode transfer
- Partitions up to a size of 8 GByte under DOS
- 50-pin Centronics plug for connection to external devices
- Not possible to boot from connected SCSI devices.

The software-driven installation (EZ-SCSI 4.0) should not present problems. If your system runs under Windows 3.1, *CardWizard Pro* has to be installed to enable the slot to be recognized at all (*CardWizard Pro* generates acoustic signals which enable the user to check if the card is recognized). Next, EZ-SCSI may be installed.

Windows 95 (not Windows NT) enables so-called *Hot Plugging*, which means that it is allowed to remove or insert a card while the system runs (obviously, not during a SCSI data transfer).

To do so, you click on the small symbol and select the command *Stop APA-1640 SlimSCSI Host Adapter*. Next, you may remove the card from the slot.

Drivers are available for all Windows platforms and OS/2.

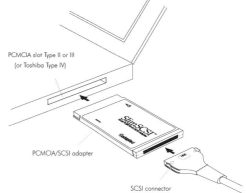

Figure 3.60.

Illustrating the individual connections when using a PCMCIA/SCSI adapter.

3. SCSI in practice

SCSI via the printer port

When a SCSI adapter is connected to a PC via the parallel printer port, the data throughput between SCSI system and CPU is totally dependent on the performance of the parallel interface.

- If the port allows uni-directional links only, the data transfer is limited to 75 kByte/s for read access, and 150 kByte/s for write access.
- If a bi-directional port is involved, a transfer speed of 260 kByte/s may be achieved for read and write access.
- The most fortunate case is that of a computer having an EPP connection (*Enhanced Parallel Port*) which allows read and write access operations to run at a speed of 1 MByte/s.

The parallel-to-SCSI adapter type *APA-385A* from Adaptec can handle all variants, and automatically adjusts the fastest possible data transfer.

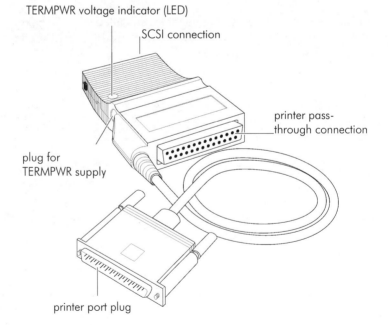

Figure 3.61. Connections on an APA-358A from Adaptec.

3.8 External SCSI adapters

The APA-358A has a passive termination at the adapter side (it is not possible to connect 'internal' devices). The powering of TERMPWR line from the adapter side is not provided here, either. Adaptec does, however, have a solution in store for the (not so rare) case that none of the connected devices is capable of taking over the supply of the TERMPWR line. An optionally available power supply (Figure 3.61) may be connected to the TERMPWR socket, and so arranges the supply. This unit is a useful extension because without it you would always have to ensure that the device which powers the TERMPWR line is switched on. In case the user is not sure if one of his devices is powering the TERMPWR line, Adaptec has an LED which lights when the supply voltage is available. By the way, this LED does not enable you to detect that the line is being powered (illegally) by several devices.

The printer port is through-connected on the SCSI adapter, so that SCSI adapter and printer may be operated on one and the same interface.

In addition to the ASPI manager, the kit contains a hard disk and CD-ROM driver including MSCDEX.EXE and a driver for HP scanners. The ASPI manager may be manually adapted to the parallel printer port by adding the extension /Mmn to the ASPI manager in the CONFIG.SYS files. *Auto Detect Mode*, is, however, the normal setting. If the (used) parallel port occupies address 3BCh, it may not be used in fast EPP mode. This mode is only possible for addresses 278h and 378h.

The adapter controls the data transfer via PIO mode, and its ID number is set to 7. A rare feature may be offered by the APA-385A — it supports Fast-SCSI timing, though in asynchronous mode only. This is insignificant for the transfer via the printer port, which does not go faster than 1 MByte/s in any case, i.e., it would not be able to profit from synchronous data transfer, and is unable to profit from Fast-SCSI timing.

4. SCSI-3 — an outlook

Although SCSI-3 has not been officially published as an ANSI standard, manufacturers already adorn their products with the addition *SCSI-3 Version*. Enquiring about the advantages of a likewise labelled device, the answers are usually evasive, and rarely precise.

The point to note is that some 'jewels' from the SCSI-3 set have already made it into the existing SCSI world: SCAM and Ultra-SCSI are just two examples. It is, however, also accurate to note that there will be no such thing as **the** SCSI-3 standard the future.

Rather, SCSI-3 will be a platform to which individual elements like Fibre Channel, SSA, Fire Wire may be attached, SCSI-3 only ensuring error-free communication between individual modules.

4.1 Modular structure

The platform consists of so-called *SCSI-3 Primary Commands* (SPC), which every device should be able to handle. Additionally, there are more specific commands for the individual device classes. So far, the following have been defined:

- *SCSI-3 Block Commands (SBC)* for hard disks and removables
- *SCSI-3 Stream Commands (SSC)* for tape drives
- *SCSI-3 Graphic Commands (SGC)* for all graphics input and output devices like scanners, plotters and printers
- *SCSI-3 Medium-Changer Commands (SMC)* for medium changers
- *SCSI-3 Controller Commands (SCC)* for SCSI adapters
- *SCSI-3 Multimedia Commands (MMC)* for CD-ROM drives, CD writers, etc.

If additions are required, new groups may be set up, the process defining the *SCSI Architecture Model (SAM)*. Grave differences with

4. SCSI-3 — an outlook

SCSI-2 are not noticed yet, but then the SCSI-2 standard offered commands for all device classes, which were completed by commands specially aimed at individual device classes.

Also known is the interface between SCSI devices on the one hand and the operating system (or application) on the other — *SCSI-3 Common Access Method (CAM-3)*.

Whether or not it will succeed in nibbling at the popularity of the ASPI manager remains to be seen. The advantages of CAM are the platform-overlapping definition of the interface, while ASPI sticks to a distinction between different operating systems.

Below the *Primary Commands* (see Figure 4.1) are the protocols of the interfaces via which the different bus types (parallel, 8-, 16- and 32-bit SCSI, serial, Fibre Channel, SSA, etc.) are allowed to dock into the system.

The usual, parallel SCSI bus (8-bit, 16-bit, ...) employs the *SCSI-3 Interlocked Protocol (SPI)*, an extension of the SCSI-2 protocol, each of the serial standards having its own protocol.

- Parallel SCSI Bus — *SCSI-3 Interlocked protocol (SIP)*
- Fibre Channel — *SCSI-3 Fibre Channel Protocol (FCP)*
- Fire Wire — *SCSI-3 Serial Bus Protocol (SBP)*
- SSA — *SCSI-3 Serial Storage Protocol (SSP)*

A general-purpose protocol was defined to facilitate access of other interfaces to SCSI:

- General-purpose interface — *Generic Packetized protocol (GPP)*.

This concept of 'docked bus systems' allows unified command sets to be employed for the entire system. No specific commands for Fire Wire or SSA — the primary commands and the device class commands apply equally to 8-bit SCSI and Fibre Channel.

On the one hand, this ensures the absolutely essential downward compatibility, on the other, it opens the gate for future developments.

It remains fair to ask, however, where the advantages of the new bus systems are with respect to the traditional, parallel SCSI bus.

4.1 Modular structure

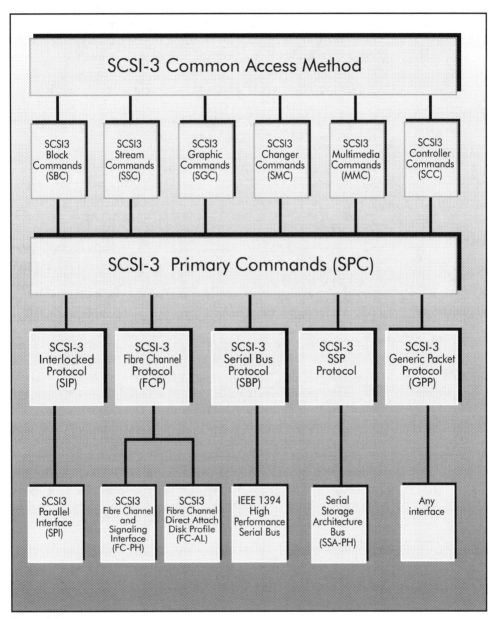

Figure 4.1.
Individual modules of the SCSI-3 architecture. The simple integration of newly arriving equipment is guaranteed, as is downward compatibility with the known and tested.

4. SCSI-3 — an outlook

As you probably know, data are transmitted either synchronously or asynchronously on the traditional SCSI bus. All commands, messages and sense data are, however, transmitted in asynchronous mode only. This *command overhead* becomes rather annoying when relatively small amounts of data are being transmitted. In a worst-case situation, more control words have to be transmitted than data! Regarding the new bus systems, the aim is to reduce this overhead.

Secondly, the allowable cable length of the parallel SCSI bus is limited to 3 or 1.5 m (Ultra-SCSI) at high transmission rates, so that external SCSI devices have to be placed quite close to the computer. Although it is possible to resort to differential SCSI (see Appendix) and be able to use cables with a length of up to 25 m, the use of serial bus systems presents a much more cost-effective solution. In this way, single-ended devices may continue to be used and the cables are certainly cheaper relating tho the length (copper). After all, 68-wire cables are replaced by 6 to 9-wire types when serial links are applied.

The third and most important reason is the aim to devise a bus system between computer and peripheral that is capable of handling much higher data rates than considered normal these days. The target level is currently about 100 MByte/s, which is not matched by the traditional SCSI bus in spite of Wide-SCSI and Fast-20 timing (a part of SCSI-3, see Figure 4.1). Although it is possible, in theory, to achieve 80 Mbyte/s using 32-bit Wide-SCSI and Fast-SCSI timing (Ultra-SCSI), that does require the use of expensive cables while the bus length is limited to 1.5 m. Taking into account the command overhead, a mere 60 MByte/s remains for the net data rate. It appears unlikely that the transfer rate offered by the parallel bus system can be improved again to a considerable degree.

In the foreseeable future, professional computer systems will, however, require data transfer rates of 100 Mbyte and more, and SCSI in particular is in wide use on high-end systems. To make sure that these systems remain faithful to SCSI, the system has to show perspective. In other words, there may not be a clearly distinguishable border indicating 'this is the limit, nothing beyond this point'.

From these general reflections you may already deduce that *Fibre Channel*, *Fire Wire*, *SSA* or whatever their names may be, are bus systems that extend SCSI systems for the professional high-end range. It is doubtful whether they will ever make it 'down' to the desktop PC range.

Consequently, the modular structure of SCSI-3 enables the bus system to be tailored to each power class. As before, the traditional parallel SCSI bus is used on systems in which the current transfer rates are sufficient (this will apply to most PCs). SCSI-3 will, therefore, not replace SCSI-2, but complete it, because the most important new features of SCSI-3 concerning the parallel bus (SCAM, Ultra-SCSI) have been introduced already.

That leaves us with the question in how far a change from SCSI-2 to SCSI-3 is noticeable if you look at the parallel bus only.

You may count on it that the command sets have been extended in some computers, while the *Mode Select* parameters and the error recognition codes will be revised and completed. The phase sequence is to be completed with a *Setup* phase after the *Select* or *Reselect* phase. Such transitions already occurred with the change from SCSI-1 to SCSI-2. If, in future, a SCSI-3 device having an extended command set is operated on the bus, and the additional commands are to be used, then a SCSI-3 adapter is required, else, the new device will operate with the reduced SCSI-2 command set.

It is, however, possible to operate SCSI-2 devices on a SCSI-3 adapter.

The cables may give rise to problems. As you may know from Chapter 3, only P- and Q-cables should be used under SCSI-3, so that it may be necessary to employ adapters for the connection of SCSI-2 devices. However, all of this is only speculation — the standard has not been published yet, and there is no SCSI-3 device yet, or, indeed, a matching adapter.

4. 2 Fibre Channel

The name of this bus standard already indicates that the designers have considered the use of fibre-optic cables. However, other alternatives may be used also.

4. SCSI-3 — an outlook

Let's start with the low-end version. The simplest cable type that may be used is the screened, twisted-pair cable. This allows transfer rates of up to 25 MByte/s, however, in contrast with the parallel SCSI bus, at a length of up to 50 m. If longer cables and/or higher transfer rates are required, coaxial or fibre-optic cables must be used. Because the interface logic remains the same, it is then perfectly legitimate to apply different cable types within an FC system. Coax cables enable up to 100 Mbyte/s to be achieved at cable lengths of up to 25 m, or 12.5 MByte/s at 100 m, and all values in between. Fibre-optic cables take us into a totally different dimension, allowing distances of up to 10 km to be covered at a maximum data transfer rate of 100 MByte/s.

As shown in Figure 4.2, Fibre Channel supports other I/O protocols besides the FCP/SCSI protocol, and even pure networking protocols (**I/O:** *IPI = Intelligent Peripheral Interface; HIPPI = High Performance Parallel Interface; SBCCS = Single Byte Command Code Set (IBM)*, **Networking:** *IP = Internet Protocol; ATM = Asynchronous Transfer Mode*). So, Fibre Channel is able to link, for example, disk arrays to networks via one and the same interface.

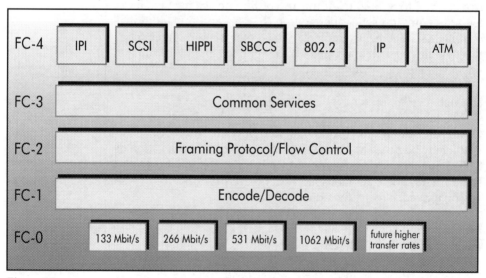

Figure 4.2.
The individual layers FC0 through FC4 of the Fibre Channel protocol allow a clear division to be made between physical and logic elements.

4.2 Fibre Channel

FC-0, the lowest layer, prescribes the physical properties for all connections, transmitters and receivers. Depending on the combination of applied elements, different transfer rates are achieved.

FC-1 defines the 8- to 10-bit ending/decoding schemes to be used, which are applied to combine clock rates and net data in such a way that they may be sent across the serial line as a *DC-balanced signal* (identical low and high levels), making error correction easier.

FC-2 describes the transport mechanisms employed by Fibre Channel, for example, how data is bundled into packets, and transmitted via the serial line. All packets of an individual transfer are normally numbered from 0 to n, so that the receiver is not only able to note that a packet is missing, but also identify it.

FC-3 embraces the definition of different transmission and information functions, for example, *Striping*, or the bundling of wires with the aim of increasing the bandwidth. The information functions may be likened to those supplied by a telephone exchange, which has only one number but several lines. If the phone rings, it is always on a line which happens to be free.

FC-4 enables the integration of various I/O standards (including FCP/SCSI) and network standards.

The most powerful connection structure of the Fibre Channel is the so-called *Interconnection Fabric*, which makes it possible to establish links between widely different parties in the connection structure, as in a telephone system.

The basic possibilities are elucidated by the illustration in Figure 4.3.

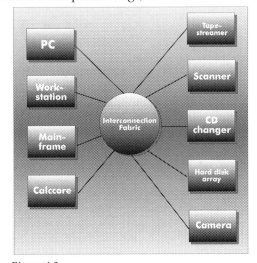

Figure 4.3.
Connections galore, made possible by the Fibre Channel's Interconnection Fabric.

4. SCSI-3 — an outlook

Depending whether fixed or variable block lengths are used, 94 to 98 per cent of the overall transfer is available on the Fibre Channel for data transfers, in other words, the overhead is 6 per cent at the most.

It consists of a Start frame, a Frame header, a Checksum and an End frame, 36 bytes in all, which are added, in the worst case, to a minimum block length of 512 bytes (hard disks).

If you start complaining now, claiming that SCSI is an I/O system and network connectivity is not yet feasible, though it holds a promise for the future, rest assured, there's also a slimmed-down version of the Fibre Channel.

FC-AL

FC-AL (*AL* for Arbitrated Loop) is a version of the Fibre Channel designed for local use. It is capable if driving up to 127 ports arranged in a loop. Each port consists of a read input and a write output. The data transfer follows the following pattern: data arrive at the read input of the first port; next, the device checks if it should accept the data, if not, they are fed out again via the write output. They reach the input of the next port, and so on.

To be able to transmit data itself, the device has to acquire control of the bus with the aid of arbitration (analogous to the parallel bus). A simultaneous transfer between different devices, as on the 'large' Fibre Channel, is not possible with FC-AL. Because of the necessary arbitration, control activities increase, and with them, control overhead. Under realistic operating conditions, however, this will not exceed 15 to 30 per cent.

The theoretically achievable transfer rates are identical to those of the 'large' Fibre Channel. Because we are talking about a local bus system, it is unlikely for fibre optics to be used. Using twisted-pair cables, up to 25 MByte/s may be achieved at a cable length of 50 m. Coaxial cables achieve 100 Mbyte/s at 10 m, or 50 MByte/s at 20 m.

As a matter of course, FC-AL may also be connected to the Fibre Channel via an Interconnection Fabric.

4. 3 SSA

A competitor of FC-AL is the SSA (*Serial Storage Architecture*) bus developed by IBM. Like FC-AL, it uses a point-to-point link. It does, however, have two read and write channels available, so that more than 127 ports may exchange data. The performance of the bus is strongly dependent on the ingenuity applied to distributing the read and write requests (*RAID* sends its regards). Each individual channel has a transfer rate of 20 MByte/s, so that 80 Mbyte/s may be achieved if the most favourable distribution is applied (each port having fifty-fifty read/write access distribution). The bus is, therefore, aimed at RAID systems, because in a single-disk only 20 MByte/s may be achieved (as only one channel may be used). This is easily matched by SCSI systems.

Like FC-AL, the SSA bus supports Hot Plugging, automatic configuration and error recognition using the 8B/10B code (*DC-balanced signal*). In contrast with FC-AL, however, SSA can make do without arbitration. Instead, data is packaged in so-called *tokens*. Each token (one per channel) has a special identification which enables the device to decide between processing received data or just sending them on. If a device wishes to transmit data, these are appended to a free token. Special tokens allow the normal bus traffic to be held up, and inject their own data. The priority assigned to the device determines how these options may be employed.

A disadvantage of SSA as compared with FC-AL may become apparent with multimedia applications, which sometimes require the data flow to be constant. For this, FC-AL offers *isochronous* data transfer, which uses a constant block size. SSA is unable to guarantee a constant data flow for each mode of operation.

Fire Wire

Although the Fire Wire interface developed by Apple was the first technically realised serial interface for the SCSI bus, it failed to get wide acceptance to date. Too few manufacturers are prepared to support this interface, even Apple remains in the wings. Fire Wire enables a data throughput of up to 50 MByte/s, and also manages isochronous data transfer, which seems to make it fit for multimedia application.

4. SCSI-3 — an outlook

Parallel or serial?

It is too early to say which of the runners in the race for the title 'serial SCSI interface' is going to be victorious. In desktop systems, they are not likely to become significant in the foreseeable future, the parallel SCSI bus with its Wide-, Fast- and Ultra-SCSI options still reigning in this area.

To the average user and his desktop PC system, the only real benefits that remain from the SCSI-3 innovation trend are the Plug & Play functions provided by SCAM, the shortened timing values of Fast-SCSI (Fast-20) and the certainty that SCSI is fully armed for future transfer rates.

5. Troubleshooting guide

This chapter provides answers to SCSI-related question that have general interest. The sources I have drawn upon are various FAQ (frequently asked questions) lists found on the Internet, readers' letters found in computer magazines, particularly those in *c't* magazine, and my own experience.

? *Is it possible to give two computers access to a single SCSI hard disk drive?*

As far as the SCSI protocol is concerned, yes, because that allows several SCSI adapters to be used on the bus. The response of operating systems is, however, a different kettle of fish. If any operating system would control the hard disk as if it were the sole user, data loss will easily come about because one computer overwrites another's data.

If, on the other hand, such a link is employed to move data from one computer to another in a simple way, and the user takes precautions to prevent the two computers from accessing the disk drive simultaneously, then it may be a useful solution, provided, of course, you know what you are doing.

? *Can I use an ohmmeter or a voltmeter to check if the SCSI bus is properly terminated?*

Yes, although a distinction should be made between active and passive termination.

With passive terminators, an ohmmeter is used to measure the resistance between any signal line and the TERMPWR line (computer and all SCSI devices switched off, of course). As shown in Figure 3.3, the resistance between a signal line and TERMPWR should be 220Ω, or 330Ω between the signal line and ground. Figure 5.1 indicates that our measurement is the result of 17 parallel-

5. Troubleshooting guide

*Figure 5.1.
Result of an ohmmeter measurement between TERMPWR and a signal line.*

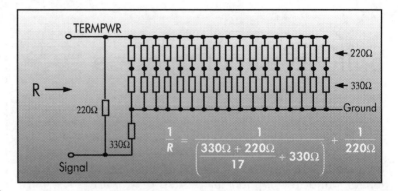

connected 550Ω resistors in series with 330Ω, this resistor network being in parallel with a 220Ω resistor between the signal line and TERMPWR.

*Figure 5.2.
A SCSI Sniffer for the 50-way Centronics-style connector. The SCSI link is connected-through, allowing measurement wires to be connected also.*

In the case of active terminators, an ohmmeter is not very useful because of the voltage regulator that sits between the signal wire and the TERMPWR line, and between TERMPWR and ground (Figure 3.7). Here, only a voltage measurement, carried out with the system up and running, produces meaningful results. The meter readings are then best taken during the Bus-Free phase. For simple test purposes, a so-called *SCSI Sniffer* (Figure 5.2) may be more suitable. The LEDs light in response to the voltage levels on the signal wires, allowing the TERMPWR voltage to be monitored in an easy way. Those of you with access to a storage oscilloscope, are, of course, even better equipped.

The measurements must be carried out separately for the internal and external bus.

Questions & answers

? How do I know if an unmarked terminator is active or passive?

Here, too, the ohmmeter comes to your rescue. Measure between the TERMPWR connection and a ground wire.

- A value of about 30Ω is measured if you are holding a passive terminator for single-ended devices, irrespective of the meter polarization.
- With active terminators, much higher resistances are measured. These values also change considerably when the test probes are swapped.
- With passive terminators for differential SCSI, the value is about 45Ω irrespective of the meter polarization.

? What is FPT (Forced Perfect Termination) all about?

This is a special type of termination where two diodes are used to prevent signal surges. They act as clipping devices, simply cutting off the voltage when it becomes too high. The diodes are not connected to ground and +5V, however, but to +1V and +3V rails supplied by voltage regulators (Figure 5.4). Because of these levels, surges are clipped earlier, and more effectively suppressed.

Forced Perfect Termination results in a better signal shape than passive termination. It does, however, draw too much current (10mA more than allowed), and should, therefore, not replace active termination as in Figure 3.7 at high transfer rates.

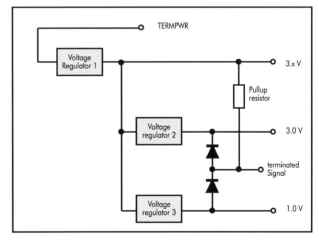

Figure 5.3.
Basic schematic of Forced Perfect Termination.

257

? *My SCSI system suffers from instability, and it often locks up with hard disk access.*

This may have several causes. Because the system runs, in principle, and all devices are recognized, a faulty adapter installation may be ruled out. It is recommended to run a systematic check on the bus.

- Are the ID numbers properly allocated? In particular, there should be no duplicate numbers.
- Check if all connections are secure.
- Is the bus terminated properly (also check the termination on the SCSI adapter)?
- Has the parity check been switched on or off, uniformly on all devices?
- Is the TERMPWR voltage available (allowable range: 4.25V to 5.25V)? This may be checked with a voltmeter or a SCSI Sniffer. A cable terminator with through connections is very useful when measuring the voltages, because the bus is then properly terminated while all individual lines are easily accessible for measuring (Figure 5.4).

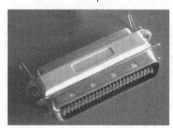

Figure 5.4.
Terminator with through-connected contacts.

- Is the TERMPWR line powered by the host adapter only, and the supply of this line switched off on all connected devices?
- Switch off all external devices to see if the error hides at the internal or the external side.
- Disconnect all devices from the bus at the suspect side, and reconnect them one by one to see which device (or device link) is causing trouble.
- Above all, check the external cables for noise susceptibility (possibly, there are wire interruptions, or the cables are very sensitive to noise).
- At the external side, use short cables only.
- Exchange passive terminators against active ones.
- Use an oscilloscope to measure the noise level on the TERMPWR voltage (perform this measurement with the bus as active as possible). If the noise level exceeds about 200mV, connect a 1μF capacitor between the TERMPWR line and ground.

Questions & answers

I hope that the system will eventually work trouble-free. If not, the fault may hide on the SCSI adapter itself.

? *My computer hangs as soon as I attempt to load the device drivers into high memory (operating system, DOS; SCSI adapter from Adaptec).*

With Bus master adapters, the ASPIxDOS manager must be loaded before the memory manager, i.e., in CONFIG.SYS, the entry

`DEVICE=...\ASPIxDOS.SYS`

must precede the line

`DEVICE=...\EMM386.SYS` (or similar).

? *The BIOS on my SCSI adapter may be switched off. Are there configurations in which this is useful?*

The BIOS should be switched off whenever boot devices are not present. This is the case when, for example, only a CD-ROM drive and/or a tape streamer are connected to the SCSI bus. The boot operation of the PC is accellerated by this measure, because the SCSI BIOS does not have to look for bootable devices.

? *I get the error message "Too many block devices" as soon as ASPIDISK.SYS is launched on my PC.*

You are obviously attempting to run too many SCSI devices. The error message indicates that insufficient drive letters are available. Modify the entry LASTDRIVE = ... in the CONFIG.SYS file.

Adapters from Symbios Logic report "Insufficient drive letters" in this case.

? *Should I continue to use ASPI or CAM drivers under Windows 95?*

No, you should remove all Real Mode drivers from the CONFIG.SYS file, and use the drivers integrated into Windows. For

5. Troubleshooting guide

example, Windows 95 uses two DLL files to installs its own ASPI platform, on to which ASPI drivers may build.

? *My SCSI hard disk has to be re-formatted. Which formatting utility do I use for this purpose?*

As a rule, SCSI adapters come with so-called low-level formatting utilities such as SCSIFMT or AFDISK. Next, the hard disk has to be partitioned using FDISK.

? *Windows 95 does not recognize all installed SCSI devices. If I use the ASPI driver, everything functions okay.*

All device drivers are automatically loaded under Windows 95. However, if Windows 95 fails to find a suitable driver for a particular device, this is not installed at all. Contact the device vendor or manufacturer and ask for a Windows 95 driver.

? *My system contains a SCSI and an IDE hard disk. Under DOS, the IDE disk boots as usual. Under OS/2, however, the PC start up from the SCSI hard disk. I am using a PCI/SCSI adapter with an NCR chip.*

Normally, OS/2 also follows the order: IDE before SCSI. The NCR driver OS2CAM.ADD is, however, known to cause the effect you describe if it is loaded before IBM1S506.ADD. Put OS2CAM.ADD at the end of your CONFIG.SYS file.

? *Is it possible to use a DDS DAT streamer to record DAT audio cassettes, and play them on an audio DAT recorder?*

That will not work. DDS DAT has an additional error detection layer, which is managed transparently by SCSI. On an audio tape, the audio bits would be interspersed with check bits, which are meaningless to the audio DAT player.

Questions & answers

? *Is there a simple way to disable SCSI devices, allowing me to determine, just as with external devices by switching off their mains power and rebooting, which devices are active on the bus?*

Switching off the mains voltage to internal devices is invariably tied to screwdriver activities and plug pulling, and, therefore, not very practical. SCSI does, however, offer the commands *Stop Unit* and *Start Unit* (see theory chapter) which, although it is not possible to disconnected devices from the mains, do allow them to be at least de-activated. Hard disk drives stop spinning as soon as a *Stop Unit* command is received. This is particularly helpful to reduce the noise level.

Using the ModePage Editor *DSP220* from DEC (available on the companion CD-ROM), you are able to generate, among others, *Start/Stop Unit* commands. The Editor runs under DOS and builds on ASPI, so that it should be compatible with just about any SCSI adapter, provided an ASPI manager is loaded.

The device to which the SCSI commands pertain has to be selected from the Select/Show sub-menu.

Careful, though! You should know what you are doing. Firstly, this editor allows other commands to be generated, too, which may cause a SCSI device to lose track, or garble the data it holds.

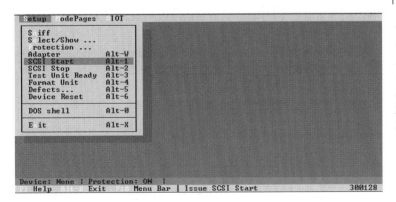

Figure 5.5. Menu shown by DEC's ModePage Editor. This program may be found on the companion CD-ROM under the name DSP220.EXE.

5. Troubleshooting guide

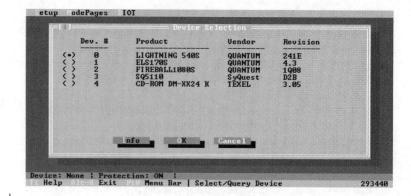

Figure 5.6.
The device to which the command relates has zo be picked from the Select/Show sub-menu.

Let's construct an example in which a number of mishaps come together:

- the DSP220.EXE file is on the boot drive;
- the boot drive is switched off using Stop Unit, and you leave the ModePage Editor.
- next, you are unable to access the boot disk and, as a result, the Editor.
- most SCSI adapters generate a Start Unit command during the boot procedure, so that the disk starts to spin again. With others, this command may be disabled, or has to be switched on again separately. This is done via the setup routine which, alas, is located on the boot disk...

So, the Modepage Editor may give you extra headaches if you do not think beforehand and are unable to carry out the functions of individual commands (see theory chapter).

Figure 5.7.
Supported Mode-Page functions of a hard disk drive.

262

Less susceptible to errors is, however, the function that allows the ModePage functions supported by a certain device to be viewed, enabling you to see if the drive unit supports cache functions, or, in the negative case, if a CD-ROM drive of a certain type (see Figure 5.8) does not recognize *Disconnect/Reselect* functions (called -/Reconnect here).

A corresponding utility for the Macintosh which offers additional, useful, functions is called *Mt.Everyhting (Version 1.1)*. It may be found in the Mac subdirectory on the companion CD-ROM. This tool enables hard disks to be post-mounted, i.e., also when no driver was installed because no medium was inserted when the system was started.

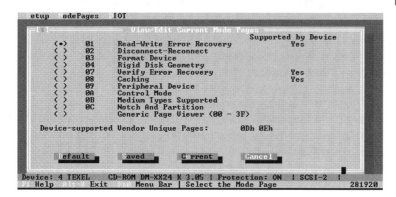

Figure 5.8. ModePage functions of a CD-ROM drive. AS you can see, Disconnect/Reselect is not supported.

? My SCSI hard disk achieves considerably lower data throughput with writing than with reading. How come?

The data transfer rate during write access operations may be increased by using the write cache on a the drive (if available). If a SCSI adapter from Adaptec is used, EZ-SCSI 4.0 may help you on (see Chapter 3 — *EZ-SCSI 4.0 — 32-bit access*). For all other adapters (including Adaptec types), Seagate's *ASPI-WCE.EXE* utility may be used, the program may be found in the companion CD-ROM. WCE stands for *Write Cache Enable*.

The command line to type from the DOS prompt is:

```
ASPI-WCE x
```

263

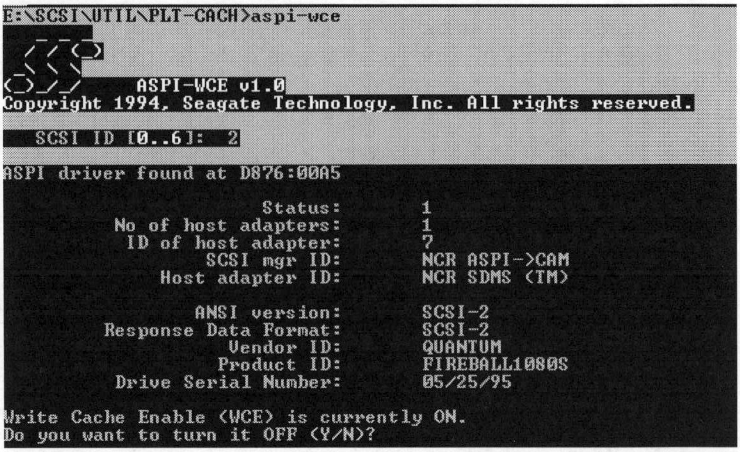

5. Troubleshooting guide

Figure 5.9.
Status report supplied by Seagate's ASPI-WCE.EXE utility.

```
E:\SCSI\UTIL\PLT-CACH>aspi-wce

         ASPI-WCE v1.0
Copyright 1994, Seagate Technology, Inc. All rights reserved.

  SCSI ID [0..6]:   2

ASPI driver found at D876:00A5
                     Status:  1
         No of host adapters:  1
          ID of host adapter:  7
                SCSI mgr ID:   NCR ASPI->CAM
             Host adapter ID:  NCR SDMS (TM)

                ANSI version:  SCSI-2
        Response Data Format:  SCSI-2
                  Vendor ID:   QUANTUM
                 Product ID:   FIREBALL1080S
          Drive Serial Number: 05/25/95

Write Cache Enable (WCE) is currently ON.
Do you want to turn it OFF (Y/N)?
```

where *x* stands for the SCSI ID used to address the hard disk. The utility supplies a status report and asks you whether the write cache settings have to be changed. (Figure 5.9).

The only condition for successful use of this utility is that an ASPI manager has been loaded. As shown in Figure 5.9, the method applied by NCR/Symbios Logic of building an ASPI driver on a CAM platform also works fine.

? *I have purchased an IBM hard disk, and use it in conjunction with an NCR adapter. The data transfer is, however, not anywhere near the value stated by the manufacturer. How come?*

Although the SDMS BIOS of the NCR adapter is capable of controlling a hard disk without the help of a driver, the disk then operates using asynchronous data transfer. A DOSCAM driver must be installed to switch over to synchronous mode. The MINICAM driver does not help you in this case, because it does not support synchronous mode either (see Chapter 3, *Symbios 8250A — CAM drivers*)

Questions & answers

? *My AHA-2940 is recognized spot-on by the BIOS, and may be configured using SCSI Select. EZ-SCSI, however, indicates that it is unable to find an adapter.*

This may be caused by a memory conflict. The ROM BIOS of the adapter is located in the upper memory range, which is also used for UMBs by memory managers. Normally, the ROM BIOS is recognized by the memory manager. If, however, EMM386 was installed with an Include statement, it may happen that the memory ranges are overwritten by UMB RAM. If that is the case, an Exclude statement should be used with EMM386 to stop it from using this particular memory range.

During the start-up process, the DOSCAM drivers (NCR/Symbios Logic) report the start address of the SCSI BIOS. If this equals, for example, E800, a corresponding Exclude entry X=E800-EFFF must be inserted in the CONFIG.SYS file.

? *I am using a hard disk and a CD-ROM drive from NEC in my PCI computer which has an NCR SCSI adapter. If I load the DOSCSAM driver, it does not always find the CD-ROM driver. In some cases, the whole system even 'hangs'.*

Looks like there is a problem with the supply of the TERMPWR line. By default, the NCR adapter supplies the termination voltage. Unfortunately, NEC CD-ROM drives are shipped with the TERMPWR suppply on. This is not allowable according to the SCSI standard. Switch off the TERMPWR supply on the NEC CD-ROM drive.

? *I have a SCSI-1 host adapter, and I would like to use it with a SCSI-2 CD-ROM drive.*

This is possible thanks to the downward compatibility of SCSI. Note, however, that the SCSI-2 device is then restricted to SCSI-1 functionality.

6. Appendix

6.1 Differential SCSI

Single-ended SCSI, using signal levels between 0 and 5 Volt, is the most commonly applied signal standard on the SCSI bus. Figure 6.1 shows the ideal signal levels and the thresholds for a logic 0 and a logic 1 to be recognized. In practice, you will never encounter such clean signal shapes as suggested in Figure 6.1.

Noise and cross-talk to other lines cause interference to be superimposed on the SCSI signal, while parasitic capacitance and junction resistances may be held responsible for the reduced edge steepness.

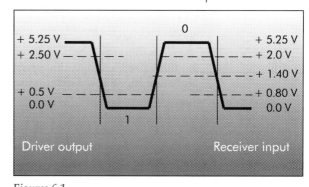

Because SCSI does not operate with a fixed clock rate, using preset signal positions and delays, problems occur as a result of interference levels, particularly if reduced timing is applied (Fast-SCSI and Fast-20 SCSI). If the interference level is too high, the signals have not settled before the allowable delay. Alternatively, the level of foreign signals superimposed on the signals may be so high that it becomes impossible to discriminate between a 0 and a 1 around the threshold level. Active terminators and shortened bus cables are then possible solutions. Shorter cables in particular do, however, form a severe restriction on the use of external devices. After all, not every device matches the

Figures 6.1.
Ideal signal sequence with single-ended SCSI. The allowable voltage tolerances for voltage outputs (drivers) are shown to the left, the levels at the receiver side, to the right. The switching threshold at the receiver side is at 1.4 V. When the instantaneous signal level is below the threshold, a 0 is detected, while a 1 is detected if the threshold is exceeded.

6. Appendix

small size of a tape streamer, allowing it to be placed close to the computer.

Ever since the introduction of SCSI-2, and not just recently with the arrival of a serial bus system applied under SCSI-3, differential SCSI may be the solution for professional computer systems which are 'spread' over several rooms, and in addition pose higher requirements as regards data transmission.

Signal levels

Whereas, under single-ended SCSI, any signal is conveyed over its own line carrying the name of the signal, and back again via a ground line, under differential SCSI, an individual wire pair is provided which carries a balanced (symmetrical) signal. Using this system, a signal level is recognized as a logic 1 if the +signal is more positive than the -signal. Conversely, a 0 is detected if the -signal is more positive than the +signal. A push-pull driver is used to generate this signal form.

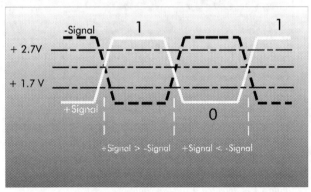

Figure 6.2.
Level sequence of the symmetrical signal as used with differential SCSI. A high degree of noise immunity is achieved by virtue of the large differences between +signal and -signal.

The level difference between a logic 1 and a logic 0 being greater than 1 Volt, the effect of external interference levels is much smaller than with the Single Ended Bus.

Because of this, the maximum cable length for a SCSI bus using differential signals is 25 metres. Even if Fast-SCSI timing is applied, the maximum allowable bus length need not be shortened.

These bus lengths, do, however, call for the use of *Twisted Pair Cables*. A sensor signal (DiffSens) applied to one cable line serves

6.1 Differential SCSI

to protect the push-pull drivers in differential devices. Because the same pin is at ground with single-ended cables, the sensor signal is short-circuited as soon as a single-ended device is (erroneously) connected to a differential bus (for wire assignment see section 6.2).

Because of its higher noise immunity, the differential SCSI bus can make do, in most cases, with a passive termination. Because neither the +signal nor the -signal are at ground potential, three resistors are then present between the TERMPWR line and ground (see Figure 6.3).

Although the pinning is also different (see section 6.2), the same cables may be used for single-ended and differential SCSI (forgetting about the length for a moment). After all, the connectors are identical. If that is an advantage, it comes at the price of a number of risks. Although the *DiffSens* helps to avoid damage to the drivers when the a single-ended device is erroneously connected to a differential bus, it is also true that a terminator, because of the identical connection type, also fits on a single-ended bus. Such a cable termination will not be successful there.

Fortunately, because of the limited spread of differential SCSI, there is only a small risk of terminators for the different signal types being interchanged. Should you hit upon problems related to the two types of terminator, you may find it helpful to read the answer to the question "*How do I know if an unmarked terminator ...*" in Chapter 5.

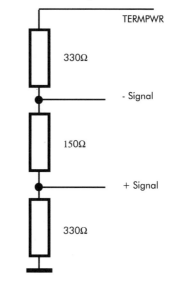

Figure 6.3.
Basic circuit diagram of a passive terminator for differential SCSI.

The advantages and disadvantages of differential SCSI may be summarized as follows:

- ☺ 25 metres maximum cable length.
- ☺ Passive termination, also with Fast-SCSI.
- ☺ High transmission security thanks to higher noise level distance (possibly, this may enable a higher transmission speed

269

6. Appendix

to be reached, because single-ended devices fall back to asynchronous transmission at too high interference levels).

☺ The same cables may be used as with single-ended SCSI.

☹ Differential and single-ended devices may not be mixed on the same bus.

☹ Terminators are specific for either bus type, and may not be interchanged. Although they look the same, single-ended and differential terminators are vastly different electrically.

☹ The large majority of SCSI devices (particularly those having a low transfer performance) is not available in a differential version.

☹ SCSI adapters for this bus type in particular are more expensive.

Note:
Adapters are available that promise conversion of the differential bus to single-ended, or the other way around. Such adapters are not foreseen in the SCSI standard, and I am not inclined to endeavour a statement as regards their reliability.

6.2 SCSI cable connector pinouts

Signal name	Plug pin		Signal name
GROUND	1	2	-DB(0)
GROUND	3	4	-DB(1)
GROUND	5	6	-DB(2)
GROUND	7	8	-DB(3)
GROUND	9	10	-DB(4)
GROUND	11	12	-DB(5)
GROUND	13	14	-DB(6)
GROUND	15	16	-DB(7)
GROUND	17	18	-DB(P)
GROUND	19	20	GROUND
GROUND	21	22	GROUND
RESERVED	23	24	RESERVED
FREE	25	26	TERMPWR
RESERVED	27	28	RESERVED
GROUND	29	30	GROUND
GROUND	31	32	-ATN
GROUND	33	34	GROUND
GROUND	35	36	-BSY
GROUND	37	38	-ACK
GROUND	39	40	-RST
GROUND	41	42	-MSG
GROUND	43	44	-SEL
GROUND	45	46	-C/D
GROUND	47	48	-REQ
GROUND	49	50	-I/O

Figure 6.4.
Pin assignment on A-cable, single-ended SCSI, internal pinheader connection

Signal name	Plug pin		Signal name
GROUND	1	26	-DB(0)
GROUND	2	27	-DB(1)
GROUND	3	28	-DB(2)
GROUND	4	29	-DB(3)
GROUND	5	30	-DB(4)
GROUND	6	31	-DB(5)
GROUND	7	32	-DB(6)
GROUND	8	33	-DB(7)
GROUND	9	34	-DB(P)
GROUND	10	35	GROUND
GROUND	11	36	GROUND
RESERVED	12	37	RESERVED
FREE	13	38	TERMPWR
RESERVED	14	39	RESERVED
GROUND	15	40	GROUND
GROUND	16	41	-ATN
GROUND	17	42	GROUND
GROUND	18	43	-BSY
GROUND	19	44	-ACK
GROUND	20	45	-RST
GROUND	21	46	-MSG
GROUND	22	47	-SEL
GROUND	23	48	-C/D
GROUND	24	49	-REQ
GROUND	25	50	-I/O

Figure 6.5.
Pin assignment on A-cable, single-ended SCSI, external connection.

6. Appendix

Figure 6.6.
Pin assignment on A-cable, differential SCSI, internal pin-header connection.

Signal name	Plug pin		Signal name
GROUND	1	2	GROUND
+DB(0)	3	4	-DB(0)
+DB(1)	5	6	-DB(1)
+DB(2)	7	8	-DB(2)
+DB(3)	9	10	-DB(3)
+DB(4)	11	12	-DB(4)
+DB(5)	13	14	-DB(5)
+DB(6)	15	16	-DB(6)
+DB(7)	17	18	-DB(7)
+DB(P)	19	20	-DB(P)
DIFFSENS	21	22	GROUND
RESERVED	23	24	RESERVED
TERMPWR	25	26	TERMPWR
RESERVED	27	28	RESERVED
+ATN	29	30	-ATN
GROUND	31	32	GROUND
+BSY	33	34	-BSY
+ACK	35	36	-ACK
+RST	37	38	-RST
+MSG	39	40	-MSG
+SEL	41	42	-SEL
+C/D	43	44	-C/D
+REQ	45	46	-REQ
+I/O	47	48	-I/O
GROUND	49	50	GROUND

Figure 6.7.
Pin assignment on A-cable, differential SCSI, external connection.

Signal name	Plug pin		Signal name
GROUND	1	26	GROUND
+DB(0)	2	27	-DB(0)
+DB(1)	3	28	-DB(1)
+DB(2)	4	29	-DB(2)
+DB(3)	5	30	-DB(3)
+DB(4)	6	31	-DB(4)
+DB(5)	7	32	-DB(5)
+DB(6)	8	33	-DB(6)
+DB(7)	9	34	-DB(7)
+DB(P)	10	35	-DB(P)
DIFFSENS	11	36	GROUND
RESERVED	12	37	RESERVED
TERMPWR	13	38	TERMPWR
RESERVED	14	39	RESERVED
+ATN	15	40	-ATN
GROUND	16	41	GROUND
+BSY	17	42	-BSY
+ACK	18	43	-ACK
+RST	19	44	-RST
+MSG	20	45	-MSG
+SEL	21	46	-SEL
+C/D	22	47	-C/D
+REQ	23	48	-REQ
+I/O	24	49	-I/O
GROUND	25	50	GROUND

6.2 SCSI cable connector pinouts

Signal name	Pin number		Signal name
GROUND	1	35	GROUND
GROUND	2	36	-DB 8
GROUND	3	37	-DB 9
GROUND	4	38	-DB 10
GROUND	5	39	-DB 11
GROUND	6	40	-DB 12
GROUND	7	41	-DB 13
GROUND	8	42	-DB 14
GROUND	9	43	-DB 15
GROUND	10	44	-DB P1
GROUND	11	45	-ACKB
GROUND	12	46	GROUND
GROUND	13	47	-REQB
GROUND	14	48	-DB 16
GROUND	15	49	-DB 17
GROUND	16	50	-DB 18
TERMPWR	17	51	TERMPWR
TERMPWR	18	52	TERMPWR
GROUND	19	53	-DB 19
GROUND	20	54	-DB 20
GROUND	21	55	-DB 21
GROUND	22	56	-DB 22
GROUND	23	57	-DB 23
GROUND	24	58	-DB P2
GROUND	25	59	-DB 24
GROUND	26	60	-DB 25
GROUND	27	61	-DB 26
GROUND	28	62	-DB 27
GROUND	29	63	-DB 28
GROUND	30	64	-DB 29
GROUND	31	65	-DB 30
GROUND	32	66	-DB 31
GROUND	33	67	-DB P3
GROUND	34	68	GROUND

Figure 6.8.
Pin assignment on B-cable for Wide-SCSI, single-ended version. This pinout is hardly found any more, at least not on newer devices. Instead of it, the P-Cable is used.

6. Appendix

Figure 6.9.
Pin assignment on B-cable for Wide-SCSI, differential version.

Signal name	Pin number		Signal name
GROUND	1	35	GROUND
+ DB 8	2	36	-DB 8
+ DB 9	3	37	-DB 9
+ DB 10	4	38	-DB 10
+ DB 11	5	39	-DB 11
+ DB 12	6	40	-DB 12
+ DB 13	7	41	-DB 13
+ DB 14	8	42	-DB 14
+ DB 15	9	43	-DB 15
+ DB P1	10	44	-DB P1
+ ACKB	11	45	-ACKB
GROUND	12	46	DIFFSENS
+ REQ	13	47	-REQB
+ DB 16	14	48	-DB 16
+ DB 17	15	49	-DB 17
+ DB 18	16	50	-DB 18
TERMPWR	17	51	TERMPWR
TERMPWR	18	52	TERMPWR
+ DB 19	19	53	-DB 19
+ DB 20	20	54	-DB 20
+ DB 21	21	55	-DB 21
+ DB 22	22	56	-DB 22
+ DB 23	23	57	-DB 23
+ DB P2	24	58	-DB P2
+ DB 24	25	59	-DB 24
+ DB 25	26	60	-DB 25
+ DB 26	27	61	-DB 26
+ DB 27	28	62	-DB 27
+ DB 28	29	63	-DB 28
+ DB 29	30	64	-DB 29
+ DB 30	31	65	-DB 30
+ DB 31	32	66	-DB 31
+ DB P3	33	67	-DB P3
GROUND	34	68	GROUND

6.2 SCSI cable connector pinouts

Signal name	Pin number		Signal name
GROUND	1	35	-DB 12
GROUND	2	36	-DB 13
GROUND	3	37	-DB 14
GROUND	4	38	-DB 15
GROUND	5	39	-DB P1
GROUND	6	40	-DB 0
GROUND	7	41	-DB 1
GROUND	8	42	-DB 2
GROUND	9	43	-DB 3
GROUND	10	44	-DB 4
GROUND	11	45	-DB 5
GROUND	12	46	-DB 6
GROUND	13	47	-DB 7
GROUND	14	48	-DB P
GROUND	15	49	GROUND
GROUND	16	50	GROUND
TERMPWR	17	51	TERMPWR
TERMPWR	18	52	TERMPWR
RESERVED	19	53	RESERVED
GROUND	20	54	GROUND
GROUND	21	55	-ATN
GROUND	22	56	GROUND
GROUND	23	57	-BSY
GROUND	24	58	-ACK
GROUND	25	59	-RST
GROUND	26	60	-MSG
GROUND	27	61	-SEL
GROUND	28	62	-C/D
GROUND	29	63	-REQ
GROUND	30	64	-I/O
GROUND	31	65	-DB 8
GROUND	32	66	-DB 9
GROUND	33	67	-DB 10
GROUND	34	68	-DB 11

Figure 6.10.
Pin assignment on P-cable for Wide-SCSI, single-ended version.

6. Appendix

Figure 6.11.
Pin assignment on Q-cable for 32-bit Wide-SCSI, single-ended version.

Signal name	Pin number		Signal name
GROUND	1	35	-DB 28
GROUND	2	36	-DB 29
GROUND	3	37	-DB 30
GROUND	4	38	-DB 31
GROUND	5	39	-DB P3
GROUND	6	40	-DB 16
GROUND	7	41	-DB 17
GROUND	8	42	-DB 18
GROUND	9	43	-DB 19
GROUND	10	44	-DB 20
GROUND	11	45	-DB 21
GROUND	12	46	-DB 22
GROUND	13	47	-DB 23
GROUND	14	48	-DB P2
GROUND	15	49	GROUND
GROUND	16	50	GROUND
TERMPWR	17	51	TERMPWR
TERMPWR	18	52	TERMPWR
RESERVED	19	53	RESERVED
GROUND	20	54	GROUND
GROUND	21	55	terminated
GROUND	22	56	GROUND
GROUND	23	57	terminated
GROUND	24	58	-ACKQ
GROUND	25	59	terminated
GROUND	26	60	terminated
GROUND	27	61	terminated
GROUND	28	62	terminated
GROUND	29	63	-REQQ
GROUND	30	64	terminated
GROUND	31	65	-DB 24
GROUND	32	66	-DB 25
GROUND	33	67	-DB 26
GROUND	34	68	-DB 27

6.3 The Companion CD-ROM

Apart from the full SCSI-2 specification, the CD-ROM that comes with this book contains various utilities, most of which are mentioned in this book.

Programming aids and additional explanations of the ASPI and CAM interface are also supplied, as well as working drafts relating to the current state of affairs regarding SCSI-3.

Furthermore, the CD-ROM contains a plethora of drivers, updates and installation assistance files for numerous SCSI adapters (the emphasis being on Adaptec and NCR/Symbios). These are the most recent drivers and updates which were available at the time of printing this book.

Readers having access to the Internet may in addition find WWW and FTP locators used by the companies mentioned to offer a powerful update service for drivers and utilities.

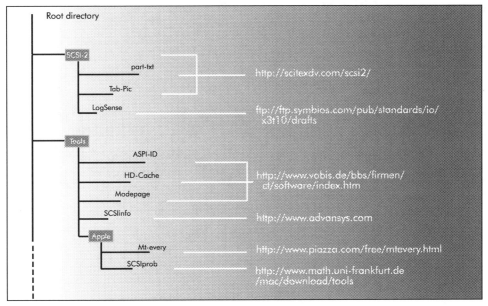

Figure 6.12. Directory structure of the CD-ROM.

6. Appendix

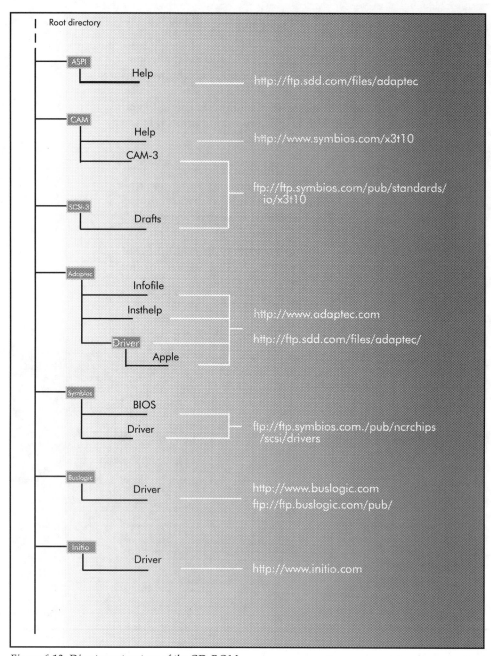

Figure 6.12. Directory structure of the CD-ROM.

Glossary

An overview of some important terms used in this book.

ANSI
American National Standards Institute. A USA based standardization authority which is also responsible for the SCSI standard.

ASPI
Advanced Programming Interface. A standardized software interface which acts as a bridge between SCSI adapter and SCSI device driver. Device drivers are tailored to the interface, not to the SCSI adapter.

Asynchronous Transfer
A transfer method under SCSI which requires every data set to be immediately acknowledged by a handshake. This is the only possible transfer mode under SCSI-1. Commands, code and status reports are always transmitted asynchronously under SCSI.

BIOS
Basic Input/Output System. An EPROM loaded with a program containing elementary computer functions, and required when the system is started. A SCSI BIOS contains basic functions for the operation of the SCSI bus, and is generally located on the SCSI adapter circuit board.

Booting (from *bootstrapping*)
Describes the start-up phase of a computer, when the operating system is loaded.

Bus Mastering
An extremely powerful method for data transfer between host adapter and the computer's main memory, which bypasses the CPU, leaving it available for handling other functions.

Bus Phase
The individual activities on the SCSI bus are completed in individual steps, called 'phases'. Each phase is determined by currently present signals, and describes the current state of the system.

Cache
A fast add-on memory, which buffers frequently used data, thus allowing fast access with repeated operations. A cache memory may be set up in the working memory of the CPU with the aid of software (software cache). Alternatively, it may be installed at the hardware level, and so support a specific peripheral device. The write cache of a SCSI hard disk acts as an intermediate storage device for writable data, to enable the bus to be freed again almost instantly. Next, the data are first fed out of the cache memory and written on to the hard disk, allowing the next action to be started on the SCSI bus.

CAM
Common Access Method. Also a standard software interface functioning as bridge between SCSI adapter and SCSI device driver. It makes the SCSI functions accessible to the device driver.

CCS
Common Command Set. A set of commands defined under SCSI-1 with the aim of unifying the commands for SCSI hard disks. Extended to all device classes under SCSI-2.

Delay
Because SCSI does not operate with a fixed clock rate, transmissions on the SCSI bus require minimum values for the signal duration and the periods between two signals. Delays are purposely inserted between signals.

Device Class
Ten device classes were defined under SCSI to enable unified command sets to be used for SCSI devices having identical or similar characteristics. Device manufactures have to put their products into a certain device class, and adapt the commands accordingly.

Differential SCSI
Slightly exotic version of SCSI which is applied mainly in larger computer systems. The term differential characterizes the signal form which consists of a +signal and a -signal. The advantages of differential SCSI include a larger allowable cable length, and higher noise immunity.

Glossary

Device driver
A driver which establishes the link between operating system or application, and the software interface (CAM or ASPI) for a particular device type (e.g., a CD-ROM drive).

DMA
Direct Memory Access. A method which allows the data flow to or from the working memory to be controlled by hardware. The hardware component (SCSI adapter) has to be set up to be able to use this mode.

EPROM
Erasable Programmable Read-Only Memory. A memory component in which important data are stored. If an EPROM is exposed to intense ultraviolet light for several minutes, the stored data are cleared, and the component may be re-programmed.

Fast-SCSI
A SCSI mode which operates with reduced timing (shorter signal and delay times), thus increasing the data throughput with synchronous data transfer from 5 Mbyte/s to a maximum of 10 MByte/s.

Glitch
A short interfering pulse on a line. Glitches are caused by cross-talk or other noise sources.

Handshake
SCSI employs a transmission method in which each data transmission has to be requested, and its reception has to be acknowledged. The request is accomplished with a *REQ* signal, the acknowledgement, with an *ACK* signal. Consequently, the term *REQ/ACK* handshaking is used. In asynchronous transmission mode, each individual *REQ* signal is confirmed by an *ACK* signal. If synchronous transmission is employed, both requests and acknowledgements may be joined.

Host Adapter
In general, any computer which is equipped with an extension card is called a 'host'. A SCSI adapter integrated into a computer

forms the link between the SCSI bus and the host computer. It is, therefore, called a host adapter.

ID number
Each SCSI device is marked by an ID (identification) number. The presettable ID number equals the address of the device. The SCSI ID also arranges the priority allocation on the SCSI bus.

Initiator
A SCSI device which initiates an I/O process.

Int13h
A software interrupt for I/O functions, which is used by the DOS operating system. SCSI adapters have to translate Int13 commands into SCSI commands.

I/O
Input/Output. There exist I/O functions, I/O programs and I/O hardware components which serve to organise the inputting and outputting of data within the computer system.

ISA
Industry Standard Association. A PC extension bus whose slots may receive plug-in type extension cards, allowing these cards to be integrated into the computer system. The ISA bus exists from the earliest years of the PC. Its width of 8 bits allowing a data throughput of just 3 MByte/s, the ISA bus is not considered powerful any more.

Logic Block
The basic data unit which is stored on most SCSI devices. Fixed and variable blocks exist.

LUN
Logical Unit. Up to 7 SCSI devices may be operated on the bus (15 or 31 under Wide-SCSI). In this arrangement, each device may contain up to eight sub-units.

Master
If a connection exists between two SCSI devices, the device in control is referred to as the Master; the other device is then the Slave.

Glossary

Multitasking
A multitasking system allows parallel execution of several commands (the emphasis is on executing here).

Multithreading
A situation in which a system (a host adapter) has to process several commands for two or more devices (the emphasis is on processing here).

Nexus
Name for a logic connection between an Initiator and a Target (or its sub-unit or queue).

Overhead
The term overhead refers to the part of a transmission that serves to control the system. The complete transmission of commands, codes and messages (in synchronous mode) on the SCSI bus is called SCSI overhead. If, for instance, a connection allows up to 10 Mbyte/s to be conveyed, and the overhead is 20 per cent, then only 8 Mbyte/s remains to convey user data.

PIO
Programmed Input/Output. A data transmission process whereby the CPU transfers data to or from the working memory via the computer's I/O ports. This mode is useful with operating systems which are not multitasking-capable (for example, DOS).

Plug & Play
A standard propagated by Microsoft aimed at virtually automatic configuration of peripheral hardware.

RAID
Redundant Array of Inexpensive Disks. Concepts aimed at increasing data security and data throughput of hard disk arrays. There are six RAID levels, RAID 0 through RAID 5, which increase the data throughput or the data security in different ways, or aim at achieving a compromise between the two targets.

SCAM
SCSI Configures AutoMatically (also called Auto Magically). A specification which allows Plug & Play functions to be installed under SCSI. SCAM functions include automatic ID number allocation, 'automatic' termination of the internal bus side, and the recognition of the Mapping algorithm used with hard disks.

SCSI Device
Any device having a SCSI interface and which may, therefore, be connected to the SCSI bus. It may be a Target or an Initiator; in other words, a SCSI adapter is also a SCSI device.

Synchronous Transfer
Using synchronous transfer under SCSI, several data packets may be transmitted one after another without the need of immediate acknowledgement via a handshake. The handshaking acknowledgment is subsequently performed in one go. In this way, the SCSI overhead may be reduced, as well as propagation losses. Under SCSI, only pure data transfers may be completed in synchronous mode.

Tagged Queues
SCSI devices supporting tagged queues (from SCSI-2 onwards) may concatenate a number of commands in a queue, and arrange their processing in the order that appears to be the most favourable in view of the achievable transfer performance.

Target
A SCSI device which is the object of an I/O process, and is being addressed by an Initiator.

Termination
To prevent undesirable signal reflections, wires or cables carrying high-frequency signals have to be terminated at their ends by a matching resistance. i.e, whose value equals the characteristic impedance of the wire or cable. Under SCSI, the resistor used for this purpose is called a terminator, and the operation is called termination.

Twisted Pair
A cable whose wires are twisted pairwise. According to the SCSI

standard, the wires in a *twisted pair* cable should be distributed such that a twisted pair always consist of a signal wire and a ground wire.

VDS
Virtual DMA Services. During DMA transfers, memory areas are addressed directly, i.e., the CPU is bypassed. In Virtual 86 program mode (available on Intel 80386 CPUs and up), however, memory ranges may be stashed, so that direct DMA transfers may lead to addressing errors. In that case, a VDS driver ensures that direct memory addressing is accomplished without errors.

Wide-SCSI
The transfer rate on a bus may be increased by raising the clock rate (or reducing the signal and delay times), but also by widening the bus. The Wide-SCSI standard allows the original bus width to be doubled (16-bit) or quadrupled (32-bit). As a result, the achievable data rates are also doubled or quadrupled.

Index

A

A-Cable 145
Abort Tag Message 62
acknowledge signal 50
addressing 20
Apple Tech Info Library 149
Arbitration Phase 32, 33
archiving medium 119
ASPI 155
ASPI interface 177
ASPI Manager 158
ATN signal 50
audio commands 112
audio tracks 115

B

B-Cable 145, 152
base address 142
BIOS 160
BIOS update 165
block address 94
block gap 106
block, logic 11, 69, 105
Boot Target ID 139
BOP, begin of partition 103
bridge controller 109
bus allocation 35
bus clock 39
Bus-Free Phase 32

C

cable impedance 39
cable length 23, 39, 47, 137
caddy 224
CAM 155
CAM Control Blocks (CCB) 156
CD-ROM devices 111

287

Index

Centronics plug 148
Check Condition 87, 115
checksum 140
CMOS components 165
Command class 66
Command Descriptor Block 67
command link 72
command overhead 248
Command Phase 32
Command structure 68
Common Command Set (CCS) 15, 22
Communication devices 124
control line 31
cross-talk suppression 39

D

Data buffer 106
Data cache 95
Data Phase 32
Data Pointer 58, 65
data transfer rate 101
data transfer, isochronous 253
data transfer, asynchronous 43
data transfer, synchronous 32, 44, 47, 59
Data-In Phase 43
Data-Out Phase 43
DC-balanced signal 251
DDS-1 235
Defect Management 116
defect memory range 95
delay 35
device class 23
Device driver 173
DiffSens signal 268
Direct Access Devices 93, 98, 199
Direct Memory Access 167
Disconnect/Reselect 172, 199, 203, 216
DMA channel 142, 228
DMA transfer 206
downward compatibility 265

drive letter 174, 223, 259
dummy drive 223

E
edge triggering. 168
Entry Point. 159
EOP, end of partition. 103
EPP connection 242
Erase gap 106
error cause 79
error correction 112, 225
even parity 140
EW, early warning. 103
EZ-SCSI 4.0. 191

F
Fast-20-SCSI 200
Fast-40-SCSI 200
Fast-SCSI. 23, 45
Fast-SCSI timing 47, 131, 137, 145, 196
FC-AL 252
Fibre Channel 219
File marks 106
Flag bit. 72
Flash BIOS 179
flatcables 144
Forced Perfect Termination. 257
Format Unit 119

G
glitches. 39, 150
grown defect list 95

H
handshake 19, 42
handshaking overhead. 19
HD connection. 148
HD socket 202
helical-scan recording 102

I

HIGH byte	208
Hot Plugging	241

I

I/O processor	131
ID number	20, 48, 138, 174, 181, 201, 207, 211, 240
Identify Message	55, 61, 64
Incorrect Initiator Connection	62
Initiator	17, 29, 34, 205
Initiator messages	55
Inquiry Data	79
Int13h	163
Interconnection Fabric	251
Interleave factor	100
interrupt	142
Interrupt Sharing	188
IRQ allocation	184
IRQ line	228
Isolation Mode	211

J

jump marks	105

L

level triggering	168, 193
linear mapping	98
Link bit	72
LOW byte	208
Low Level Format	178, 260
LUN	58, 60, 176, 231

M

Mapping	93
Master functions	29
Master/Slave assignment	29
Medium Changer devices	123, 125, 227
Medium Scan	119
memory block	93
memory manager	259
Message Phase	33

Message-In Phases... 41
Mode parameters... 67
Mode Select command... 101
Mode Sense command... 115
Modepage Editor... 262
MSCDEX.EXE... 175
Multi-Master Ability... 29
multi-threading... 172

N
network... 125
nexus... 63

O
on-board adapter... 169
operating parameters... 75
Operation Code... 68
Optical Memory devices... 115
OS/2... 185
OS/2-ADD Specification... 185

P
P-Cable... 145, 152
Page code... 78
parallel track recording... 102
Parity checking... 140
Partitions... 103
Pass-Through Functionality... 187
PCI bus... 131
PCI master... 167
PCI slave adapter... 190
PCI slot... 165
PCMCIA... 239
permission to access... 34
Phase-Change Technology... 229
PIO mode... 167, 190
pixel... 121
Plug & Play... 209
Plug connection... 152
Pre-Fetch... 95, 97
Primary Commands (SPC)... 245

primary defect list 95
priority. 35, 139
priority sequence. 20
propagation loss 44, 47

Q

QIC standard 235
QIC-80 streamer 234
Queue Tag message 62

R

RAID. 204, 209
reflection factor. 132
Removable media 97, 221
REQ signal 44
REQ/ACK handshake. 42, 44
REQ/ACK Offset 45, 59
Reselection Phase 32, 35, 39
Reset command. 33
Reset Condition. 50
response time. 36
RIF bit 121
rules for termination 134

S

SASI 11
SASI standard 69
SCAM 183, 209, 219
SCAM Select Sequence 211
scan range 120
scanner. 120
SCSI adapter 13, 21, 29
SCSI Architecture Model (SAM). 245
SCSI BIOS 161, 173, 181
SCSI cables. 143, 145, 147, 149, 151, 153
SCSI command 69
SCSI device. 13
 external 215, 222
SCSI ID. 139
SCSI interface. 13
SCSI Medium Changer. 123

SCSI processors.................................. 109
SCSI Reset...................................... 136
SCSI Sniffer.................................... 256
SCSI-2 connector................................ 143
SDMS system 170
Second Level Error Correction 111
Selection Abort Time 51
Selection Phase 32, 38
Send Diagnostic.................................. 87
Sense Data 74, 79
Sequential Access 101
serpentine track recording...................... 102
Session... 119
Set marks 106
Setup phase 249
shift register................................... 97
Shugart.. 11
signal form...................................... 18
SIM (SCSI Interface Module)..................... 155
Single-Ended Devices 167
SSA bus 253
status message 49
Status Phase.................................. 32, 72
storage volumes.................................. 69
Stripe Set 203
Striping 251
sub-D plug 149
Synchronous Data Transfer Request............... 59

T

tagged queue 61, 63, 172
tape marker 105
Target..................................... 17, 29, 61
Target messages.................................. 57
Target routines.............................. 61, 65
termination............................ 20, 201, 210
    passive 133
    errors 135
    power 132
    resistance 132
Terminator 29, 137, 138, 143, 150

TERMPWR 29
TERMPWR line 136, 169, 218, 240
timing value. 45, 47
track group 102
transfer mode. 19
transfer parameter 176
Transfer Phases 40
transfer rate 47, 180
TWAIN 236, 237
twisted-pair cables 146
Type-Byte Identification. 233

U

Ultra Wide-SCSI adapter 201
Ultra-SCSI. 200
Unix drivers. 189

V

Virtual 8086 Mode 171
Virtual DMA Service 171

W

Wide Data Transfer Request. 60
Wide-SCSI 23, 145, 148, 195
Wide-SCSI, 32-bit. 200
Window Descriptor. 121
Windows NT. 186, 199
Windows95 182
write cache 96, 194, 263
Write-Once devices. 119, 228

X

XPT 155

Sources used for photographs and drawings

Adaptec: page 130, 166, 167, 241, 242
Panasonic: page 230
Philips: page 229
Plextor: page 224
Polaroid: page 239

All other drawings and photographs: Ulrich Weber.